The Arms Race and Arı

sipri

Stockholm International Peace Research Institute

SIPRI is an independent institute for research into problems of peace and conflict, especially those of disarmament and arms regulation. It was established in 1966 to commemorate Sweden's 150 years of unbroken peace.

The Institute is financed by the Swedish Parliament. The staff, the Governing Board and the Scientific Council are international.

sipri

Stockholm International Peace Research Institute

Bergshamra, S-171 73 Solna, Sweden

Cable: Peaceresearch, Stockholm

Telephone: 08-55 97 00

The Arms Race and Arms Control

sipri

Stockholm International Peace Research Institute

Taylor & Francis Ltd
London

First published 1982 by Taylor & Francis Ltd
4 John Street, London WC1N 2ET

Distributed in the United States of America and Canada by
Oelgeschlager, Gunn & Hain, Inc.,
1278 Massachusetts Avenue, Harvard Square,
Cambridge, MA 02138, USA
and in Scandinavia by
Almqvist & Wiksell International,
26 Gamla Brogatan,
S-101 20 Stockholm, Sweden

ISBN 0 85066 232 X

Typeset by The Lancashire Typesetting Co. Ltd,
Bolton, Lancashire BL2 1DB
Printed and bound in the United Kingdom by
Taylor & Francis (Printers) Ltd, Rankine Road,
Basingstoke, Hampshire RG24 0PR

PREFACE

More and more people are getting worried about the arms race, about the risks of war, and in particular about the possibility that nuclear weapons might actually be used. They are beginning to see a real and present danger. Groups of concerned people are forming, across occupational and political lines, to study the problems and to find ways of letting politicians know of their concern.

This book is for those who want well-researched information on what is going on in the world of armaments and disarmament. There are negotiations at Geneva about nuclear weapons targeted on Europe. The United States and the Soviet Union give widely different assessments of the existing balance. What are the facts? The United States has authorized the production of the enhanced radiation weapon—known as the neutron bomb. What is a neutron bomb? Both the United States and the Soviet Union claim that the other side is preparing for a 'first strike'. What are the facts about this? How many nuclear weapon tests were conducted last year—and is the number rising or falling? What has happened to the negotiations for banning them? Talks have been going on at Madrid for over a year about a European Disarmament Conference. What are the issues? This book is about questions of this kind.

The material is drawn from the 1982 edition of SIPRI's Yearbook of World Armaments and Disarmament; the full contents list of which is given at the back of this book. For thirteen years now, SIPRI has published this Yearbook. It has established itself as a neutral, unbiased source of information on questions of armaments and disarmament. The purpose of this paperback version is to make the most immediately relevant parts of the material in this Yearbook available to a wider audience.

Frank Blackaby
Director

CONTENTS

The researchers were assisted by Per Berg, Ragnhild Ferm, Gunilla Herolf, Evamaria Loose-Weintraub, Elisabeth Sköns, Carol Stoltenberg-Hansen and Rita Tullberg.

Figures were drawn by Alberto Izquierdo.

ACRONYMS

ABM	Anti-ballistic missile
AGM	Air-to-ground missile
ALCM	Air-launched cruise missile
ASAT	Anti-satellite
ASBM	Air-to-surface ballistic missile
ASM	Air-to-surface missile
ASW	Anti-submarine warfare
AWACS	Airborne warning and control system
BMD	Ballistic missile defence
BW	Biological weapon
CBM	Confidence-building measure
CBW	Chemical and biological warfare
CCD	Conference of the Committee on Disarmament
CD	Committee on Disarmament
CEP	Circular error probable
CSCE	Conference on Security and Co-operation in Europe
CTB	Comprehensive test ban
CW	Chemical weapon
ENDC	Eighteen-Nation Disarmament Committee
ENMOD	Environmental modification
ERW	Enhanced radiation weapon
FOBS	Fractional orbital bombardment system
GLCM	Ground-launched cruise missile
IAEA	International Atomic Energy Agency
ICBM	Intercontinental ballistic missile
INFCE	International Fuel Cycle Evaluation
IRBM	Intermediate-range ballistic missile
ISMA	International Satellite Monitoring Agency
LRTNF	Long-range theatre nuclear force
MAD	Mutual assured destruction
MARV	Manoeuvrable re-entry vehicle
M(B)FR	Mutual (balanced) force reduction
MIRV	Multiple independently targetable re-entry vehicle
MRV	Multiple (but not independently targetable) re-entry vehicle
MURFAAMCE	Mutual reduction of forces and armaments and associated measures in Central Europe
NPT	Non-Proliferation Treaty
NWFZ	Nuclear weapon-free zone
OPANAL	Agency for the Prohibition of Nuclear Weapons in Latin America
PNE(T)	Peaceful Nuclear Explosions (Treaty)
PTB(T)	Partial Test Ban (Treaty)
RV	Re-entry vehicle
RW	Radiological weapon
SALT	Strategic Arms Limitation Talks
SAM	Surface-to-air missile
SCC	Standing Consultative Commission (US–Soviet)
SLBM	Submarine-launched ballistic missile
SLCM	Sea-launched cruise missile
SRBM	Short-range ballistic missile
TTBT	Threshold Test Ban Treaty

GLOSSARY

Anti-ballistic missile (ABM) system	Weapon system for intercepting and destroying ballistic missiles.
Anti-satellite (ASAT) system	Weapon system for destroying, damaging or disturbing the normal function of, or changing the flight trajectory of, artificial Earth satellites.
Ballistic missile	Missile which follows a ballistic trajectory (part of which is outside the Earth's atmosphere) when thrust is terminated.
Battlefield nuclear weapons	*See:* Theatre nuclear weapons.
Binary chemical weapon	A shell or other device filled with two chemicals of relatively low toxicity which mix and react while the device is being delivered to the target, the reaction product being a supertoxic chemical warfare agent, such as nerve gas.
Biological weapons (BW)	Living organisms or infective material derived from them, which are intended for use in warfare to cause disease or death in man, animals or plants, and the means of their delivery.
Chemical weapons (CW)	Chemical substances—whether gaseous, liquid or solid—which might be employed as weapons in combat because of their direct toxic effects on man, animals or plants, and the means of their delivery.
Circular error probability (CEP)	A measure of missile accuracy: the radius of a circle, centred on the target, within which 50 per cent of the weapons aimed at the target are expected to fall.
Committee on Disarmament (CD)	Multilateral arms control negotiating body, based in Geneva, which is composed of 40 states (including all the nuclear weapon powers). The CD is the successor of the Eighteen-Nation Disarmament Committee, ENDC (1962–69), and the Conference of the Committee on Disarmament, CCD (1969–78).
Conventional weapons	Weapons not having mass destruction effects. *See also:* Weapons of mass destruction.
Counterforce attack	Attack directed against military targets.
Countervalue attack	Attack directed against civilian targets.
Cruise missile	Missile which can fly at very low altitudes (and can be programmed to follow the contours of the terrain) to minimize radar detection. It can be air-, ground- or sea-launched and carry a conventional or a nuclear warhead.
Enhanced radiation weapon (ERW)	*See:* Neutron weapon.
Enriched nuclear fuel	Nuclear fuel containing more than the natural contents of fissile isotopes.
Enrichment	*See:* Uranium enrichment.
Eurostrategic weapons	*See:* Theatre nuclear weapons.
Fall-out	Particles contaminated with radioactive material as well as radioactive nuclides, descending to the Earth's surface following a nuclear explosion.

First-strike capability	Capability to destroy within a very short period of time all or a very substantial portion of an adversary's strategic nuclear forces.
Fission	Process whereby the nucleus of a heavy atom splits into lighter nuclei with the release of substantial amounts of energy. At present the most important fissionable materials are uranium-235 and plutonium-239.
Flexible response capability	Capability to react to an attack with a full range of military options, including a limited use of nuclear weapons.
Fractional orbital bombardment system (FOBS)	System capable of launching nuclear weapons into orbit and bringing them back to Earth before a full orbit is completed.
Fuel cycle	*See:* Nuclear fuel cycle.
Fusion	Process whereby light atoms, especially those of the isotopes of hydrogen—deuterium and tritium—combine to form a heavy atom with the release of very substantial amounts of energy.
Genocide	Commission of acts intended to destroy, in whole or in part, a national, ethnical, racial or religious group.
Intercontinental ballistic missile (ICBM)	Ballistic missile with a range in excess of 5 500 km.
Intermediate-range nuclear weapons	US designation for long-range and possibly medium-range theatre nuclear weapons. *See also:* Theatre nuclear weapons.
International Nuclear Fuel Cycle Evaluation (INFCE)	International study conducted in 1978–80 on ways in which supplies of nuclear material, equipment and technology and fuel cycle services can be assured in accordance with non-proliferation considerations.
Kiloton (kt)	Measure of the explosive yield of a nuclear weapon equivalent to 1 000 metric tons of trinitrotoluene (TNT) high explosive. (The bomb detonated at Hiroshima in World War II had a yield of about 12–15 kilotons.)
Launcher	Equipment which launches a missile. ICBM launchers are land-based launchers which can be either fixed or mobile. SLBM launchers are missile tubes on submarines.
Manoeuvrable re-entry vehicle (MARV)	Re-entry vehicle whose flight can be adjusted so that it may evade ballistic missile defences and/or acquire increased accuracy.
Medium-range nuclear weapons	Soviet designation for long-range theatre nuclear weapons. *See also:* Theatre nuclear weapons.
Megaton (Mt)	Measure of the explosive yield of a nuclear weapon equivalent to one million metric tons of trinitrotoluene (TNT) high explosive.
Multiple independently targetable re-entry vehicles (MIRV)	Re-entry vehicles, carried by one missile, which can be directed to separate targets.
Mutual assured destruction (MAD)	Concept of reciprocal deterrence which rests on the ability of the nuclear weapon powers to inflict intolerable damage on one another after surviving a nuclear first strike.

Mutual reduction of forces and armaments and associated measures in Central Europe (MURFAAMCE)	Subject of negotiations between NATO and the Warsaw Treaty Organization, which began in Vienna in 1973. Often referred to as mutual (balanced) force reduction (M(B)FR).
Neutron weapon	Nuclear explosive device designed to maximize radiation effects and reduce blast and thermal effects.
Nuclear fuel cycle	Series of steps involved in preparation, use and disposal of fuel for nuclear power reactors. It includes uranium ore mining, ore refining (and possibly enrichment), fabrication of fuel elements and their use in a reactor, reprocessing of spent fuel, refabricating the recovered fissile material into new fuel elements and disposal of waste products.
Nuclear weapon	Device which is capable of releasing nuclear energy in an explosive manner and which has a group of characteristics that are appropriate for use for warlike purposes.
Nuclear weapon-free zone (NWFZ)	Zone which a group of states may establish by a treaty whereby the statute of total absence of nuclear weapons to which the zone shall be subject is defined, and a system of verification and control is set up to guarantee compliance.
Peaceful nuclear explosion (PNE)	Application of a nuclear explosion for such purposes as digging canals or harbours, creating underground cavities, etc.
Plutonium separation	Reprocessing of spent reactor fuel to separate plutonium.
Radiological weapon (RW)	Device, including any weapon or equipment, other than a nuclear explosive device, specifically designed to employ radioactive material by disseminating it to cause destruction, damage or injury by means of the radiation produced by the decay of such material, as well as radioactive material, other than that produced by a nuclear explosive device, specifically designed for such use.
Re-entry vehicle (RV)	Portion of a strategic ballistic missile designed to carry a nuclear warhead and to re-enter the Earth's atmosphere in the terminal phase of the trajectory.
Second-strike capability	Ability to survive a nuclear attack and launch a retaliatory blow large enough to inflict intolerable damage on the opponent. *See also:* Mutual assured destruction.
Standing Consultative Commission (SCC)	US–Soviet consultative body established in accordance with the SALT agreements.
Strategic Arms Limitation Talks (SALT)	Negotiations between the Soviet Union and the United States, initiated in 1969, which seek to limit the strategic nuclear forces, both offensive and defensive, of both sides.
Strategic nuclear forces	ICBMs, SLBMs, ASBMs and bomber aircraft of intercontinental range.
Tactical nuclear weapons	*See:* Theatre nuclear weapons.
Terminal guidance	Guidance provided in the final, near-target phase of the flight of a missile.

Theatre nuclear weapons	Nuclear weapons of a range less than 5 500 km. Often divided into long-range—over 1 000 km (for instance, so-called Eurostrategic weapons), medium-range, and short-range—up to 200 km (also referred to as tactical or battlefield nuclear weapons).
Thermonuclear weapon	Nuclear weapon (also referred to as hydrogen weapon) in which the main part of the explosive energy released results from thermonuclear fusion reactions. The high temperatures required for such reactions are obtained with a fission explosion.
Toxins	Poisonous substances which are products of organisms but are inanimate and incapable of reproducing themselves. Some toxins may also be produced by chemical synthesis.
Uranium enrichment	The process of increasing the content of uranium-235 above that found in natural uranium, for use in reactors or nuclear explosives.
Warhead	That part of a missile, torpedo, rocket or other munition which contains the explosive or other material intended to inflict damage.
Weapons of mass destruction	Nuclear weapons and any other weapons which may produce comparable effects, such as chemical and biological weapons.
Weapon-grade material	Material with a sufficiently high concentration either of uranium-233, uranium-235 or plutonium-239 to make it suitable for a nuclear weapon.
Yield	Released nuclear explosive energy expressed as the equivalent of the energy produced by a given number of metric tons of trinitrotoluene (TNT) high explosive. *See also:* Kiloton and Megaton.

NOTE ON CONVENTIONS

The following general conventions are used in the tables:

.. Information not available

( ) Uncertain data or SIPRI estimate

– Nil or not applicable

'Billion' in all cases is used to mean thousand million.

Metric units generally apply. However, both short tons and metric tons are used and are specified where necessary. For convenience, the conversions are:

1 *metric* ton (tonne) = 1 000 kilograms = 2 205 pounds = 1.1 short tons

1 *short* ton = 2 000 pounds = 0.91 metric ton (tonne)

1 kiloton (kt) = 1 000 (metric) tons

1 megaton (Mt) = 1 000 000 (metric) tons

The dose of radiation is measured as the energy of the ionizing radiation absorbed in tissue. The unit of dose is the Gray (Gy); 1 gray = 1 joule per kilogram of tissue. Many publications still use the *rad* as the unit of dose (1 Gy = 100 rad).

1. Introduction

The purpose of this book—and of the summary in this introduction—is to review the state of world armaments and disarmament, in advance of the United Nations Second Special Session on Disarmament.

Obviously matters of armaments and disarmament are interconnected with international political events: consider, for example, the effect of the imposition of martial law in Poland on the discussions at Madrid. This book does not set out to cover political events of that kind—it would have to be double the size to do that. Developments in what has been called 'the world war industry' are proper subjects of study in their own right—the fact that there is a UN Special Session on Disarmament is evidence enough of that. Armaments are not simply the consequence of international tension: they are also a cause.

Since the First Special Session on Disarmament four years ago, things have got worse. Expenditure on military research and development is rising fast; the spread of modern weapons around the world continues unchecked. There is little impetus at the moment behind any moves for arms control, let alone disarmament. The pressure against the few arms control barriers which have been set up in the post-war period is getting stronger. It is a sign of the times that some people are beginning to talk of the present as a pre-war rather than a post-war period.

The hopeful sign is in growing public concern, particularly in some countries in both Western and Eastern Europe and particularly about nuclear weapons—a concern not simply about the nuclear weapons of one side only. Questions of disarmament are no longer matters of limited interest to a small circle. As a consequence, the major powers—in the negotiations at Geneva for example—are having to take public opinion into account. Both the US and Soviet Ministries of Defence have published popular books on the threat to peace presented by the other side. The need for unbiased information was never greater.

The short summary which follows has to be highly selective. It begins with world military expenditure, the production of conventional weapons, and the arms trade. It then looks at the growing arsenals of intercontinental nuclear weapons—and in this weapons section summarizes the material on the militarization of outer space, on the neutron bomb, and on chemical and biological warfare. The third section, on armaments and arms control, concentrates on the background to the negotiations at

Geneva, and presents some main points from a discussion of Nordic initiatives for a nuclear weapon-free zone.

I. World military expenditure, arms production and the arms trade

During the past four years, world military spending has been following an upward trend at a rate of about 3 per cent per annum (in volume). This is rather faster than in the previous four years, in spite of the deteriorating performance of the world economy. So the burden, measured as a share of the world's total output, has probably been rising. It is difficult to get a meaningful measure of the world total: for what it is worth, the current dollar figure in 1981 was about $600–650 billion.

There is no evidence of any particular change in trend in Soviet military spending: a steady rise continues. The Soviet Union outproduces the United States in its annual deliveries of a number of standard conventional weapons; that has been true for a long time. The technological lag, however, though it may be smaller than a decade ago, is still considerable, particularly in electronics. A military comparison must allow for the fact that European NATO countries have bigger military budgets than the other Warsaw Treaty Organization (WTO) countries, and that the Soviet Union also maintains a considerable military force along its border with China.

The Soviet Navy continues to improve its ocean-going capacity, with a number of new classes of ship which will give the Soviet Union a much greater peace-time 'power projection' capacity. This capacity is still inferior to that of the United States. On the other hand, the Soviet Union is much nearer than the United States to certain important existing and potential areas of confrontation—the Persian Gulf, the Middle East, Korea and Europe itself.

There has been a sharp change in trend in military spending in the United States. This already appears in the 1981 figures, where the estimated volume increase in military spending for the calendar year is 6 per cent. The new Administration's five-year plan is indeed to move military expenditure (actual outlays) on to an 8 per cent real growth path—that is, the average annual percentage change from now to 1987 implied by the figures in the 1983 budget request. This follows a substantial change in public attitudes: back in 1969, in a public opinion poll, only 8 per cent of respondents said that defence spending was too small. By 1980 the figure had risen to 49 per cent.

The rearmament programme includes a number of new strategic weapon systems (discussed in section II). Otherwise, the main objective is to increase the ability of the United States to project its power in parts of the world distant from the American continent. For the Navy, the aim is to reach a 600-ship Navy by 1987: that means the construction of 143 combat ships. For the Army and the Marines, heavy expenditure is envisaged for the Rapid Deployment Force (RDF). The decision has also been taken to resume production of chemical weapons, which had been stopped over a decade ago.

The main question mark over this programme is an economic one. If the programme is put into effect, military spending will increase its share of national output from 5.7 per cent in 1981 to 7 or 8 per cent in 1986, depending on whether there is a recovery in productivity. The future course of US military spending will, quite probably, be mainly determined by economic factors.

Whereas in the United States the trend in military expenditure has begun to accelerate, in other NATO countries (taken together) it has not. Since May 1977, when NATO countries collectively agreed to adopt an annual 3 per cent volume target increase, the rise in military spending in NATO countries other than the USA has been slightly slower than it was before. Most countries in Europe have been preoccupied with their budget deficits; finance ministers have won out over defence ministers. Many politicians saw no reason to think that the Soviet threat had suddenly become so acute as to require dramatic changes in their military spending.

The divergence between the United States and its NATO allies is likely to lead to stresses within the alliance. So, too, is the United States' development of weapons—the neutron bomb, and chemical munitions with binary agents—which only make sense if deployed in areas of possible confrontation such as Europe, but which the Europeans in general do not seem to want.

In the United Kingdom, there has been an upward change in trend—though even so a defence review has forced reductions in the Navy's surface fleet. The main source of public concern has been with the independent nuclear deterrent—first with an immensely expensive programme whose object was to try to ensure that Polaris warheads could penetrate possible future anti-ballistic-missile defences round Moscow; and secondly with the escalating cost of the future replacement of Polaris with the Trident system. In the Federal Republic of Germany military spending has not risen much in real terms, and there the major concern has been with the budget cost of the Tornado (the multi-role combat aircraft) programme; a series of upward revisions brought the 1981 cost of this programme up from DM 1 750 million to a figure of DM 3 065 million.

Japan has also been under pressure from the United States to increase its military budget, with the suggestion that it should take responsibility for defending the airspace and sea lanes up to 1 000 miles from its shores.

The suggestion has been met with a cool response. There is little public enthusiasm in Japan for more military spending. The article in the constitution which says that "land, sea and air forces, as well as other war potential, will never be maintained" still has some influence. Nevertheless, Japan ranks eighth in the world in its expenditure on Self-Defense Forces.

Military spending is moving up significantly in India and Pakistan, with substantial arms supplies from the Soviet Union and the United States respectively. After a long period of relative quiescence, Australia and New Zealand are also increasing their military budgets. This is a reaction to the general increase in world tension, rather than the perception of any new threat. The one major country where the change has been in the other direction is China. In 1981, the Chinese military budget was cut heavily. Top priority is at present being given to the improvement of the civil economy.

Arms trade

There is at present little prospect for any kind of restraint on the international trade in arms. The conventional arms transfer talks between the United States and the Soviet Union were adjourned three years ago, and have not been resumed; the European arms suppliers have shown no inclination towards restraint. International tension and economic pressure all make for bleak prospects for any restraint. The underlying trend—doubling in volume every five years—continues.

In the period 1979–81, the Soviet Union overtook the United States as the leading exporter of major weapons. This was partly because of a big increase in arms exports to India, and to countries in the Middle East and North Africa; the other reason was a decline in US exports resulting from the policy of restraint initiated by President Carter in 1977.

However, the Soviet Union still has a smaller number of customers than the United States: during 1981, it had arms deals with 28 countries, compared with 67 countries for the USA. The Soviet Union traditionally charges low prices, has favourable credit terms, and has been prepared to consider barter arrangements; however, more recently it has been looking for payment in hard currency. It is also exporting more modern equipment than before: for example, it is believed that the slow introduction of the T-72 main battle tank into service with the WTO armies is partly explained by large exports to Middle Eastern and North African countries. The Soviet Union is using arms transfers as an important instrument for maintaining and expanding its influence in the Third World. Arms transfers play a far greater role than economic aid or trade in this respect;

Figure 1. Exports of major weapons to the Third World compared with world trade, 1962–80

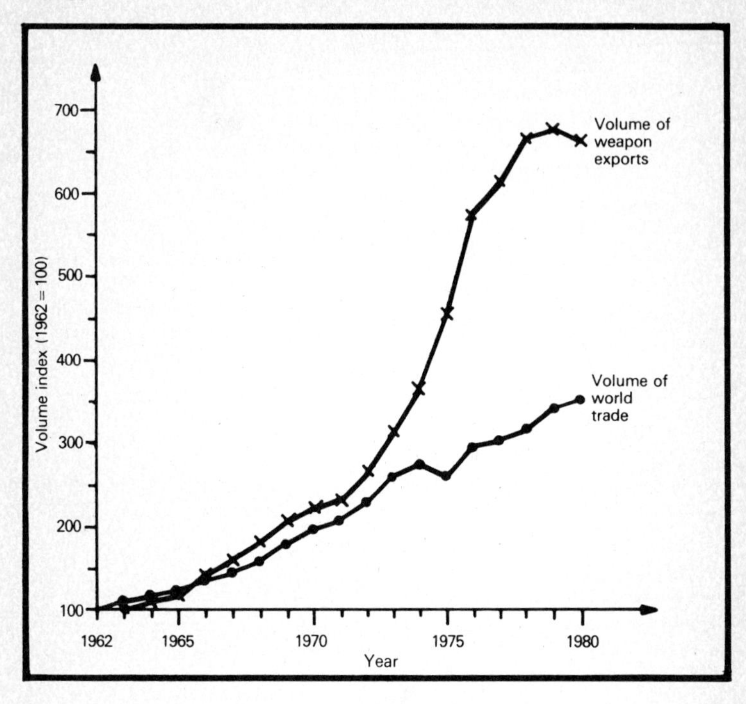

Sources: Exports of major weapons to the Third World—SIPRI data bank. World trade—United Nations *Statistical Yearbook*, *1974* and *1978*; *UN Monthly Bulletin of Statistics*, January 1982.

it is virtually the only area in which they have successfully rivalled the West.

In the United States, the policy of restraint on arms sales, which President Carter enunciated in a May 1977 directive, has now been abandoned. A new directive was issued in July 1981, which reinstates arms sales as a major instrument of foreign policy. Security assistance authorized for the fiscal year 1982 shows an increase of 30 per cent, compared with fiscal year 1981; a substantial part of that assistance consists of foreign military sales financing. Human rights issues, as embodied in Carter's 1977 directive, will not be a significant consideration. The constraint of not introducing advanced weaponry that would raise the combat capability in any given region—also in the 1977 directive—has been abandoned as well. Thus, South Korea will get an initial batch of

36 F-16s, which is an introduction into that region of weaponry of a significantly higher technological level than before.

The main events in the US arms trade in 1981 were the deals with Pakistan and Saudi Arabia; and there was the significant decision that China could, if it wished, buy 'lethal' weapons from the United States—though there are few signs of China wishing to do so at the moment. The United States negotiated a $3.2 billion five-year military and economic package with Pakistan, including 40 F-16 fighters. With Saudi Arabia, an air defence package was negotiated which is probably the largest single arms transaction of the post-war period. It includes five AWACS aircraft, six aerial refuelling tankers, 1 177 Sidewinder air-to-air missiles, and 22 ground-based radar installations. Given the historical Saudi opposition to foreign military bases on their soil, the AWACS deal is the nearest thing to a prepositioned base that the United States is likely to obtain at this stage, at least until the Saudis themselves are able to operate and maintain these systems.

West European countries have been pushing arms sales in 1981. The new French Administration does not appear to have made any change in French arms export policy: for instance, during 1981 France delivered Mirage fighters to Iraq and missile-armed attack boats to Iran. Libya also received French weapons during much of 1981. The UK has been promoting sales of the British Aerospace Hawk, Chieftain tanks and Rapier surface-to-air missiles in the Middle East; it has also lifted the embargo on arms sales to Chile. FR Germany is under some pressure to change its policy prohibiting sales to 'areas of tension'; because of this policy, a large sale of tanks and armoured vehicles to Saudi Arabia is still pending. In recent years, Italy has emerged as the world's fourth largest exporter of major weapon systems, with a policy which enables firms to export to virtually any country in the world.

Some Third World countries are now increasing their share of the arms trade with exports of domestically produced weapons. Because of the lower unit prices it is mainly other Third World countries that buy these weapons. Brazil has a booming arms industry—for example, the Engesa Company reportedly sells approximately 1 000 armoured vehicles a year to 32 countries.

The Israeli arms industry is one of the largest employers in Israel. In 1981, for instance, it sold substantial quantities of tank ammunition to a number of countries, including Switzerland; the Galil rifle was another prominent export item. There were also Israeli arms transfers during 1981 to Iran, including spare parts for US-built M-48 tanks and for F-4 Phantom fighters.

II. *Weapons*

Strategic nuclear weapons

The confrontation between the two great powers in intercontinental nuclear weapons is becoming increasingly uneasy. Each side claims that the other side is trying for some kind of first-strike capability, while declaring its own objective to be solely defensive. The United States' scenario is that the Soviet Union launches a strike which eliminates all US land-based missiles. It still has enough strategic nuclear weapons in reserve to inhibit the United States from making any reply with its submarine-launched missiles.

It is difficult to believe that any sane ruler would order a first strike of this kind—except as a pre-emptive move, in the belief that the other side was about to do the same. The risk of total catastrophe to his own country would be very large. As a realistic technological and political option, a first strike limited simply to land-based missiles lies in the realm of myth.

However, it is this myth which is being used as a rationale for the very big increases which are in prospect in strategic weapon programmes and procurement. It is also the rationale for the renewed advocacy in some strategic journals and elsewhere of a launch-on-warning system to prevent land-based missiles being caught in their silos. These missiles should be launched, without reference to the head of state, as soon as various detection devices suggest that the missiles from the other side have left their silos.

The Soviet Union is proceeding with the modernization of its land-based missiles, replacing old missiles with SS-17s, SS-18s and SS-19s. The great majority of these newer missiles are equipped with MIRVs (multiple independently targeted re-entry vehicles). The replacement of old missiles by these newer types will probably be complete by the mid-1980s. It is also anticipated that the Soviet Union will develop solid-propellant intercontinental ballistic missiles to supplement or replace some of the current liquid-propellant ones.

The most modern class of Soviet missile submarine which is operational is the Delta class, with missiles which have a range of 8 000–9 000 km. These missiles can be fired at most targets in the United States from waters close to the Soviet shore, such as the Barents Sea and the Sea of Okhotsk; thus the submarines can reduce their exposure to US anti-submarine warfare systems. In 1980, the Soviet Union launched a new, much larger strategic nuclear submarine, the *Typhoon.* This, it is believed, will carry some 20 ballistic missiles, each missile with probably 12 warheads; it will also be able to cover most targets in the United States from

Soviet home waters. It could also be deployed under the ice of the Arctic Ocean, as further protection against US anti-submarine tactics.

The Soviet Union has not taken any action for over a decade to deploy any new long-range bombers. It maintains a formidable air defence system, which it will probably wish to upgrade to deal with the US cruise missile threat.

The United States proposes to press ahead with the production of the new MX land-based intercontinental missile, which will have three times the throw-weight of the Minuteman III missile and can carry 10 warheads of about 500 kilotons each. The proposal is to deploy some 35–40 of these missiles in existing ICBM silos, and in the meantime to look at long-term basing options for this missile. One of these options—developing ballistic missile defence for the missile sites—would require the revision, or indeed possibly the abandonment, of the Anti-Ballistic Missile Treaty.

The bomber programme is the largest element in the strategic programme. Firstly, it is proposed to upgrade the B-52G and B-52H bombers so that they can carry some 3 000 cruise missiles. Secondly, 100 B-1 bombers will be built, also equipped to carry air-launched cruise missiles. Thirdly, there is an intensive research and development programme for the Advanced Technology ('Stealth') bomber. In addition to the deployment (which has begun) of the air-launched cruise missiles for the bombers, it is proposed to deploy Tomahawk cruise missiles, some of which will be nuclear-armed, on submarines and surface ships.

The first of the new Ohio-type ballistic missile submarines was commissioned in November last year; it will carry 24 Trident missiles, each with 8 100-kiloton MIRVed warheads. Eight such submarines are now being built. The development has begun of a more advanced Trident missile, the Trident II, with a longer range, and carrying more warheads. The Trident II is expected to be as accurate as a land-based ICBM. The strategic weapon programme in the USA also includes substantial expenditure on improved communications and control systems.

Nuclear explosions

Of the 49 nuclear explosions which took place in 1981, the USSR carried out 21. (Five of these were conducted outside the Soviet weapon testing sites and are therefore presumed to have served non-weapon purposes.) The USA conducted 16 nuclear weapon test explosions at the usual site in Nevada; the UK conducted 1, also in Nevada; and France conducted 11 on the atoll of Mururoa in the Pacific Ocean. China did not test at all last year.

All explosions in 1981 were carried out underground and, according to data obtained from the Hagfors Observatory in Sweden, all had a yield below or around 150 kt (the yields of the French tests were 20 kt or below).

The rate of testing in the past four years—around 50 a year—has been significantly higher than in the previous four years (1974–77 inclusive), when the average was 37 tests only. There has been no downward trend since the 1963 Partial Test Ban Treaty.

The military use of space

At least three-quarters of all satellites are used for military purposes. They are intricately connected with the development of the new strategies for nuclear weapons which have evolved with the increasing accuracy of those weapons. Satellites are used to obtain precise knowledge of the targets and their locations, and are also used in the command, control and communications systems which transmit targeting information and which direct the actions of the offensive forces.

Satellites are obviously vulnerable, and the military are concerned to find ways of improving the survivability of their own satellites, and of attacking the satellites of the potential enemy. The United States, for instance, is devoting resources to hardening the electronic components of space systems, so that they are less likely to be damaged by an electromagnetic pulse (EMP) which can be produced by the explosion of a nuclear warhead. The US Air Force has also proposed a satellite which would orbit at an altitude of around 200 000 km and would have manoeuvring capabilities. Both sides have been experimenting with methods of destroying the other side's spacecraft. The Soviet Union launched a target satellite and two interceptors during 1981. The United States is planning to begin operational testing of its anti-satellite (ASAT) system in 1983. This consists of a miniature homing vehicle which would be guided to its target by an infra-red homing device, and which could be launched from aircraft flying at an altitude of some 20 km.

Both the USA and the USSR are investigating high-energy laser and particle beams for ASAT applications. By the end of fiscal year 1981, the Department of Defense will have spent about $1.5 billion on investigations into laser weapons; even so, the United States claims that the Soviet Union is ahead in this field. The chances are that both are roughly equally advanced. During 1981, the US Air Force conducted a number of tests of its laser weapon against a Sidewinder air-to-air missile.

The neutron bomb

A neutron bomb, or an enhanced radiation weapon, is a nuclear weapon so designed that the fraction of energy released as prompt radiation is much higher than in the standard nuclear weapon, and the fraction released as blast effects is much lower. In the late 1970s, the United States developed enhanced radiation warheads for the Lance missile with a range of about 100 km and for the 203-mm artillery howitzer with a range of 29 km. In 1978, President Carter approved production of the non-nuclear but not the nuclear components for these new warheads. In the summer of 1981 President Reagan authorized, without consulting his NATO allies, the procurement and stockpiling of the complete enhanced radiation warheads. He said they would not be deployed overseas at this time. However, they are clearly intended for Europe, and would have to be moved there if they were to have any function.

The arguments presented for the new weapons are that, if used against tanks or other targets, the blast effect would be less than that of standard nuclear weapons, and the damage to civilian life and property would be less. It would therefore be more credible to the Soviet Union that they might be used, and the Soviet Union would thus be deterred from attempting a tank attack. However, the neutron bomb is not a prescription for a safe nuclear war for Europeans. First of all, significant radiation casualties could be expected over an area of 10 square kilometres for each neutron weapon used: if 1 000 such weapons were used—and that is what might be needed—there could be anything up to some 10 000 square kilometres in which Europeans would be subjected to dangerous radiation exposure.

Secondly, if they were used, the likelihood is that the Soviet Union would retaliate with nuclear weapons of its own. Once these weapons are deployed, the main danger is that the reduced blast effect might make the decision to use these weapons easier to take. The decision to fire them would probably be delegated to local commands, and hence the nuclear threshold would be lowered. Crossing the threshold from conventional weapons to the first use of any nuclear weapon would create a high risk of escalation to a nuclear war in Europe.

France is also developing neutron bombs, but a decision about their production has not yet been taken.

Laser enrichment of plutonium

There seems likely to be a rising demand in the United States in the next decade for weapon-grade plutonium. In recent years, new supplies were not needed: plutonium was recycled from obsolete nuclear weapons into new ones. Now the situation has changed. Firstly, the US rearmament programme will mean a big increase in the number of nuclear warheads deployed—certainly several thousands more, and possibly as many as 10 000 more. Secondly, plutonium is preferred to uranium in most types of nuclear warhead. Thirdly, there is competition between tritium and plutonium for the limited capacity of existing production reactors: tritium is essential for the production of all fusion weapons.

In the United States there has been a substantial research and development programme into new techniques for the enrichment of uranium. The same techniques could be used to enrich reactor-grade plutonium—the plutonium produced by the civil nuclear power industry—so that it became weapon-grade material. One such technique, which may be near the pilot plant stage, is laser enrichment. If a way is found of converting reactor-grade plutonium into weapon-grade plutonium at no great cost, the link between civil nuclear power technology and military nuclear weapon technology will be further strengthened. Moreover, the offer made by some nuclear weapon states to submit their civilian nuclear activities to international safeguards would become meaningless. Such a development would weaken the legitimacy of the Non-Proliferation Treaty and its attendant system of safeguards.

Chemical and biological warfare (CBW)

In February 1982, the President of the United States certified to Congress that it was essential to the national interest that production of chemical weapons should be resumed in the United States, after an interval of over a decade: there has been no significant production of filled poison-gas ammunitions in the United States since 1969. The world is moving to the verge of a chemical arms race that could make impossible any further strengthening of the arms control measures in this field.

The present CBW arms control arrangements rest on the 1972 Biological and Toxin Weapons Convention (BW Convention) which outlaws the development, production, stockpiling and international transfer of these weapons, and on the Geneva Protocol of 1925. The latter agreement has an important weakness: there is no international verification machinery to deal with allegations of the use of these weapons. The attempts over the past 10 years to strengthen the control over chemical weapons have so far been unsuccessful.

At the end of World War II, more than a dozen states possessed stocks of the latest chemical weapons. Now, there are only three states—France, the USA and the USSR—publicly known to possess militarily significant stocks. The current US stockpile is about 42 000 short tons of poison gas, of which about half is mustard gas, and the other half nerve gas. However, some of the nerve gas is stored in filled munitions which have either deteriorated over the years, or are obsolete. The supply of serviceable and ready-to-use poison gas munitions probably amounts to some 70 000 tons. (The tonnage of munitions exceeds the tonnage of the basic agents by a factor of about 10.) If the bulk-stored mustard and nerve gases were filled into munitions, that might add a further 200 000 tons. Most of this stockpile is held in the United States; the only two overseas stockpiles known are at Johnston Island in the Pacific and one ammunition depot in FR Germany.

US officials, not French ones, have confirmed the existence of a French chemical-weapon stockpile; it is reckoned to amount to some hundreds of tons of nerve gas.

The West has no firm information about the size of the Soviet stockpile. Soviet officials have made no direct public reference to the existence of such weapons in the Soviet Union since 1938. Current professional estimates range from less than 30 000 to more than 700 000 tons of chemical agents. This has given rise to the frequent quotation of the arithmetic mean of these two figures—350 000 agent-tons. This would correspond to about 3 300 000 tons of filled munitions—a figure so enormous as to cause grave doubts about its plausibility. The chemical agents said to be stockpiled include a variety of types of World War I and II vintage, as well as nerve gases. There is no hard evidence that the Soviet Union has been producing chemical agents or munitions during the 12 years that have elapsed since US production stopped. The Western officials who refer to a build-up have been referring, not to continued production of chemical weapons, but to the continuing build-up of anti-chemical protection that had commenced during the 1960s, coupled with the increased deployment of weapon systems capable of firing, among other things, chemical ammunition.

There are a number of sources of pressure which may sweep away the constraints which have held back a chemical arms race during the past decade. New technology makes it simpler to assimilate chemical weapons into military inventories. Chemical agents are now quick-acting, and in this respect more closely resemble conventional weapons. They are packaged in ammunition which can be used with conventional weapon systems, so that there is no longer any call for special chemical troops. The latest innovation is 'binary' nerve-gas munitions. These are shells, bombs or rocket-warheads filled, not with actual nerve gas, but with separate loadings of much less toxic chemicals adapted to mix and react together to generate nerve gas only when the munition is on its final target course. Binaries do away with the need for expensive and dangerous super-toxic chemical factories, and have sufficiently enhanced storage and handling safety to allow combat units to carry supplies with them.

It is true that adherence to the Geneva Protocol requires the military to speak only in terms of deterrence: the possession of poison gas, it is argued, is simply to deter the other side from using it. However, once these weapons have been integrated into the force structure—which indeed is necessary for them to fulfil their reputed deterrent function—the military will undoubtedly begin to look beyond deterrence to scenarios in which the no-first-use policy is abandoned.

Allegations of the actual use of chemical weapons, and other infractions, have added to the pressure against the existing arms control constraints. The allegations of the use of chemical agents in Laos, Kampuchea and Afghanistan are being examined by an expert investigatory group convened by the Secretary-General of the UN; its interim report, released in mid-November 1981, "found itself unable to reach a final conclusion as to whether or not chemical warfare agents had been used. . . . Any investigation designed to lead to definite conclusions . . . would require timely access to the areas of alleged use. Such an exercise has so far not been possible." The United States has also reiterated its accusation that the Soviet Union has acted in violation of the 1972 BW Convention. The event which stimulated the US action was an outbreak of human anthrax in 1979 in the region of Sverdlovsk—long known to be an area where anthrax is endemic. For reasons not made public, US evaluators suspect that the victims were suffering from the pulmonary rather than the intestinal form of the disease, and are unwilling to accept the Soviet explanation that it was caused by infected meat. These suspicions have been allowed to grow by the absence of any verification provisions in the Convention. There are also Cuban allegations attributing, for example, outbreaks of sugar-cane rust and blue mould of tobacco to CIA activities.

Finally, the fact that chemical disarmament negotiations were making some progress served to alert the protagonists of chemical weapons. The constant references to the existence of a chemical-warfare gap *vis-à-vis* the Soviet Union began in the summer of 1977, soon after the negotiations were joined in earnest. The arguments were presented for negotiating from a position of strength, requiring some 'bargaining-chip' chemical rearmament. On the Soviet side, there was a refusal to accept mandatory on-site inspection even of the destruction of stockpiles.

The US Department of Defense is now building a full-scale factory for making new binary nerve-gas munitions. It should be ready for operation during the fiscal year 1983, and will have a capability of 20 000 155-mm rounds per month. Next off the production line would be the 500-pound binary-VX aircraft spray-bombs (Big Eye). After that, binary warheads are being considered for a range of rockets and missiles, including the ground-launched cruise missile.

III. Arms control and disarmament

If this section were restricted to summarizing actual progress made during 1981 in arms control and disarmament, it would be short. No progress was made. There is a long list of negotiations and discussions which lie dormant (or possibly dead). Negotiations on a comprehensive test ban were adjourned in November 1980. They have not been resumed, and the US Administration has indicated that it has no interest at present in their resumption. There were talks between the United States and the Soviet Union in 1978–79 on possible control of anti-satellite systems; around the same time there was also some discussion between them on possible restraint in their sales (or gifts) of conventional arms. Neither of these discussions has been resumed. Negotiations between the United States and the Soviet Union on chemical weapons have also been in abeyance—although multilateral discussion has continued in the Committee on Disarmament.

No progress has been made at the Vienna talks—now in their ninth year—on mutual (balanced) force reductions in Europe; and at present there does not seem much chance that an agreement on the holding of a European Disarmament Conference will emerge from the Conference on Security and Co-operation in Europe in Madrid.

Perhaps the most dangerous hiatus is the absence of any negotiations on strategic nuclear weapons. The second treaty on strategic arms limitations (SALT II), laboriously negotiated over seven years, was not put to the US Senate for ratification by the previous Administration; and the present US Administration considers that the treaty is fatally flawed. However, after a year in office the new US Administration has still not agreed to a date for resuming talks. It has simply indicated that it wishes to talk about reductions rather than limitations: it has also hinted that it may have strong requirements for verification.

The one set of negotiations which has got under way is on long-range theatre nuclear forces in Europe—the LRTNF negotiations. They began in November 1981—although it is difficult to see how far they can get, unless complemented by negotiations about strategic nuclear weaponry. The following summaries begin with the LRTNF issues: a fuller summary is given at the beginning of the chapter itself, on page 3. Summaries of the Nordic proposals for a nuclear weapon-free zone (NWFZ), and of the state of negotiations at Madrid then follow. There are finally notes on militarization and arms control in Latin America, on the stage which negotiations on a comprehensive test ban had reached before they were adjourned, on the Soviet proposal for banning weapons in outer space, and the proposal for an international satellite monitoring agency.

Long-range theatre nuclear forces in Europe

Since the 1950s, the Soviet Union has had a large number of missiles with nuclear warheads targeted on Western Europe—to that extent the SS-20s do not represent an entirely new threat. The decision to replace the SS-4s and SS-5s with SS-20s may have been taken without much attention to the international political consequences. In fact, West European nations have been much concerned at the increased capabilities of these new missiles, while at the same time they were beginning to doubt whether their forward-based aircraft could continue to penetrate Soviet air defences.

In recent months, the two sides now negotiating on this matter have put forward widely different assessments of the balance. A reasoned judgement is that, whether the comparison is limited to missiles, or whether it includes aircraft as well (where the problem of deciding what to include is much more difficult), the Soviet Union appears to have a superiority in long-range theatre nuclear forces in Europe of about 2:1. Insofar as there is concern to change this particular regional balance, then obviously it is better to do so by reductions on the Soviet side than by increases on the NATO side.

If indeed new missiles were installed on the NATO side, it is a mistake to think that they would serve to re-establish the United States' 'nuclear umbrella'. There is no doubt that, if a war broke out in Europe, both major powers would attempt to keep their own homelands free from attack with nuclear weapons by initially avoiding attacks on the homeland of the other side. Thus, the new missiles, if introduced, would in all probability have a set of targets in Eastern Europe, west of the Soviet border. For if a nuclear missile fired by US forces strikes the Soviet Union, the Soviet Union would in all probability retaliate against the United States, whether the missile came from the Federal Republic of Germany or from Montana.

The Geneva negotiations, if they are to have significant success, must soon be linked with strategic arms limitation or reduction talks. Otherwise it would be too easy to negate the effect of any agreement reached—for instance by the deployment of cruise missiles on ships in northern European waters, or by introducing new missiles with ranges below 1 000 km.

By the end of 1981, the Soviet Union had some 175 SS-20 missile launchers within striking range of Europe. The number of warheads carried by 175 SS-20s is roughly the same as the number deployed on SS-4s and SS-5s before the SS-20 was introduced. The number of launchers, 175, is also roughly the same as the number now deployed by the UK and France combined. So the *status quo ante*, and a rough matching of Soviet missile systems with those of the UK and France, could be obtained by freezing the number of SS-20 launchers at their end-1981 number and eliminating all the SS-4s and SS-5s.

Nuclear weapon-free zone: Nordic initiatives

The Nordic area is not itself likely to become a source of major power conflict. However, there is an increasing risk that it may become an arena of international rivalry, with the spread of more effective nuclear war-fighting weapons to northern Europe. The proposal for a nuclear weapon-free zone (NWFZ) in the area is aimed at making it a kind of low-tension buffer zone between the major powers.

There are three main characteristics of a NWFZ: non-possession, non-deployment and non-use of nuclear weapons. All the Nordic countries have ratified the Non-Proliferation Treaty. None of them possesses or deploys nuclear weapons in peace-time, or allows them to be deployed by other countries. The main change that a NWFZ in the Nordic region would require would be that Norway and Denmark would agree not to allow the deployment of nuclear weapons on their territories in times of war. In the established definitions of a NWFZ, the prohibition applies to nuclear explosives only. There may indeed be other installations on the territories of the Nordic countries, such as sonar arrays and navigation aids for submarines, which are linked to the global nuclear-weapon strategies of the great powers; however, attempts to extend the scope of the prohibition would lead to reduced clarity.

Transit provisions would have to be regulated by the treaty, otherwise transits could be so frequent that the basic provisions could be undermined. Overflights of aircraft or cruise missiles carrying nuclear weapons would have to be prohibited: the treaty would have to negotiate provisions that cruise missiles would not be located in such a way that their trajectory would almost certainly cross zone territory.

The Soviet Union has indicated its willingness to consider "measures applying to [Soviet] territory in the region adjoining a nuclear free zone in the north of Europe". The candidates for elimination include a number of missiles in the Leningrad military district which are in all probability intended for strikes against Nordic targets; there might also be a total ban on submarine-based nuclear weapons in the Baltic Sea. Limitations near Denmark would depend probably on some progress being made in confidence-building measures over a wider European area.

A Nordic nuclear weapon-free zone could be considered as a first step towards more comprehensive measures covering the whole of Europe: alternatively, if any broader European arrangements were to be agreed first, it might be established within that framework. It could be considered together with other suggestions for confidence-building measures, such as some restrictions on anti-submarine warfare activities, or a demilitarized area along the Norwegian–Soviet border.

The case for some disengagement in northern Europe is strong, since the alternative is not the *status quo* but a big increase of military capabilities in the area. However, there is the major problem of finding a design which is acceptable to the major powers.

European Disarmament Conference

The possibility and problems of convening a European Disarmament Conference (EDC) have been discussed now for over a year at Madrid at the second review conference of the Helsinki Final Act. This conference is known as the Conférence on Security and Co-operation in Europe (CSCE). The talks have gone on for so long because, although the international atmosphere has not been propitious to an agreement, no party is anxious to take the responsibility for ending the discussions. Most states at the CSCE seem to believe in the need for convening an EDC. The differences between them concern the preparations and agenda for such a conference.

The background to the security issues dividing the CSCE may be set out by summarizing the Polish and French proposals. The Polish proposal was for a step-by-step advance from the confidence-building measures (CBMs) adopted at Helsinki towards arms control and disarmament measures, and for a conference at which a wide range of proposals could be put forward and considered. The French proposal was for a more ambitious and more detailed exchange of military information: a set of CBMs for which four criteria should be agreed before an EDC was convened. The new CBMs should be significant in military terms; they should be binding, not voluntary as heretofore; there should be appropriate verification; and they should be applicable throughout Europe from the Atlantic to the Urals. When these CBMs had been adopted and implemented, arms control and disarmament negotiations could be started.

The main controversial issue has been the area of application. The Helsinki CBMs apply to the whole of Europe, except for the Soviet Union where only the area within 250 km of the frontiers with other European states is covered by the requirement to notify manoeuvres. The Soviet Union has indicated that it might accept the extension of the area to the entire European part of the USSR, provided the western states also extended corresponding zones accordingly. The problem has been to agree how to compensate for the Soviet concession 'accordingly'. There is the possibility of establishing zones in the waters surrounding Europe where military activities would be notified; alternatively, certain military activities outside Europe which were connected with activities inside Europe could come under the notification requirement. Possibly the concessions could combine both geographical and functional requirements.

Militarization and arms control in Latin America

The application of an arms control regime to a whole inhabited continent is a new development. Latin America is unique as the first nuclear weapon-free zone on a continental scale and in a populated region, established by the 1967 Treaty of Tlatelolco. However, this has not prevented further militarization of the region in recent years.

In particular, the strength of the armed forces has almost tripled in the past two decades in Central America and the Caribbean. The militarization of this sub-region has been accompanied by an intensification of internal violence in many of these countries. In El Salvador, for example, as many as 35 000 people were killed from 1979 to the end of 1981.

South American countries have also been involved in a formidable expansion of their military potential, mainly because of the revival of inter-state border conflicts, as well as internal upheavals. Brazil and Argentina have developed significant arms industries, and are widely believed to be seeking nuclear weapon capabilities. Neither of them is a party to the Treaty of Tlatelolco.

Under these circumstances, it has not proved possible in Latin America to move on from the Treaty of Tlatelolco to further arms control measures.

Comprehensive test ban

Before the negotiations were adjourned in November 1980, a number of important points had been agreed. In particular, important advances had been made in the matter of verification. The treaty would provide for consultations to resolve questions that might arise concerning compliance, and any party would have the right to request on-site inspection for the purpose of ascertaining whether or not an event on the territory of another party was a nuclear explosion. The three negotiating parties had also agreed in principle on a number of high-quality, tamper-proof national seismic stations of agreed characteristics, to be installed on the territories of the three parties.

Although the principle of on-site inspection had been agreed, the various procedures of the inspection process had not. Another issue which may need settlement is the question of laboratory tests which could consist of extremely low-yield nuclear experiments.

There was also a point at issue on the duration of the treaty. The initial duration was to be only three years; the United States did not want to to make, in the treaty, a provision for possible extension, while the Soviet Union preferred to stipulate that the ban would continue unless the other nuclear weapon powers, not parties to the treaty, continued testing.

A comprehensive test ban ought at least to make it difficult for the nuclear weapon parties to be certain about the performance of new weapons that are developed, and to that extent would narrow one channel of arms competition among the major powers. It would also reinforce the Non-Proliferation Treaty by demonstrating that the major powers had some awareness of their legal obligation to bring the nuclear arms race to a halt.

Two proposals concerning outer space

In 1981 the Soviet Union proposed a treaty of unlimited duration, which would prohibit the stationing of weapons of any kind in outer space, including stationing on "reusable" manned space vehicles (a clear reference to the US space shuttle programme). Moreover, the parties to the treaty would undertake not to destroy, damage or disturb the normal functioning or change the flight trajectory of space objects of other states, if such objects were placed in orbit in "strict accordance" with those provisions. Compliance with the treaty would be assured by the national technical means of verification at the disposal of the parties and, when necessary, the parties would consult each other, make inquiries and provide relevant information.

This proposal is not, as yet, fully elaborated. For instance, it is not clear who would make the judgement as to whether or not objects were placed in orbit in accordance with the provisions of the treaty. Although the proposal is for a multilateral treaty, only the two major powers would have available to them "national means of verification". Further, it is probably desirable that the treaty should cover, if possible, weapons that could strike space objects from the ground or from the atmosphere.

A report has also been prepared for the United Nations General Assembly on the possibilities for setting up an international satellite monitoring agency (ISMA). The report concludes that space technology will allow observations from satellites for the verification of compliance with arms control and disarmament treaties and for monitoring crisis areas. The annual cost of an ISMA to the international community would be very much less than 1 per cent of the total yearly expenditure on armaments. There are, of course, difficult questions about the distribution of the data and the information which such an agency might acquire. There are political, organizational and financial difficulties. The idea of an ISMA could be the beginning of a multinational verification agency. However, both the USA and USSR have so far been negative, and have refused to participate in the group.

2. World military expenditure and arms production

Square-bracketed numbers, thus [1], *refer to the list of references on page* 49.

I. Introduction

World military spending has continued to rise, in real terms. Indeed the rise through the last four years—at something like 3 per cent a year—has been, if anything, rather faster than in the first half of the 1970s. This is in spite of the deteriorating performance of the world economy: world economic growth has slowed down considerably in recent years. So the burden of world military spending—measured as a share of the world's total output—has probably been rising.

It is never easy to know how best to give an impression of the size of the world's military budget. The dollar is still the standard measuring-rod used, yet it is difficult to find sensible ways of converting the military spending of Socialist countries into dollars. For what it is worth, the dollar total in 1981—at current prices—was of the order of $600–650 billion.

The main change—which has now begun—is in the United States. After a fairly long post-Viet Nam period in which US military spending was falling in real terms, it is now set to rise rapidly—and indeed has begun to do so. A formidable rearmament programme is in prospect—so formidable that some commentators believe that it will eventually be cut back because of the economic difficulties it creates.

The United States has not been successful in persuading its NATO allies to follow suit. Indeed the rise in NATO Europe's military spending has been fractionally slower in the past four years than in the previous four, and there is little in the 1982 budgets to suggest any substantial change. Japan has also been resisting US pressure to spend more than 1 per cent of the Japanese national product on defence. If these divergent trends continue—of rapidly rising military expenditure in the United States, with much smaller rates of increase in Western Europe and Japan—this is bound to create tension in the Western alliance.

There does not appear to have been any particular change in trend in Soviet military spending: high figures have continued for the output of military hardware, with a steady upward trend which does not seem to vary much from year to year.

There does seem to have been some change in the trend in India and Pakistan, where the rise in military spending has accelerated in the past two years—coupled with increased supplies of weapons from the two great powers. Australia has reacted to the general increase in world tension with a rearmament programme. In the Middle East and the Persian Gulf, perhaps the main change has been the increase in the flow of military aid; both great powers continue to supply substantial quantities of weapons. In Egypt, Oman and Somalia, the United States, in exchange for military aid, is proceeding with the construction of various base facilities.

The one major country which is following a different course is China. The Chinese view is clearly that there is no imminent threat; the military budget has been cut back substantially, in the interest of the civil economy.

The sections which follow concentrate this year on the major countries—the USSR, the USA, the main countries in Western Europe, China and Japan. (Developments in intercontinental strategic weaponry are dealt with in chapter 4.) Shorter notes follow on some other selected areas of interest. A final section briefly discusses multinational weapons production. Appendix 2A provides one small example of weapon development, as an illustration of the process—and it takes for this purpose the Maverick and Condor air-to-surface missiles.

II. The Soviet Union

There is the usual dearth of hard information from the Soviet side on the Soviet Union's military expenditure or production; as usual, the figures in this section come from Western sources. If these figures give the wrong impression, it is for the Soviet Union to correct that impression by releasing more information. In a world of satellite photography, the Soviet Union's all-pervading secrecy does little to conceal its military capabilities from the United States.

In the course of 1981, the US Department of Defense published a book entitled *Soviet Military Power*. It was widely distributed in Western Europe as part of a campaign to persuade West European audiences that there is a genuine Soviet threat. The message of the book is summed up in a preface by the US Defense Secretary as follows:

> All elements of the Soviet Armed Forces . . . continue to modernize with an unending flow of new weapon systems, tanks, missiles, ships, artillery and aircraft. The Soviet defence budget continues to grow to fund this force build-up, to fund the projection of Soviet power far from Soviet shores and to fund Soviet use of proxy forces to support revolutionary factions and conflict in an increasing threat to international stability. [1]

The general impression given by this book is of a perfected military machine: indeed it has been described as an excellent public relations document for the Red Army. There are very few references to weaknesses or inadequacies. Such a degree of perfection is unlikely. The Soviet civil economy is known to be inefficient, with low productivity and under-used capital equipment. The military sector may be more efficient than the civil sector, but it is not likely that it wholly escapes the defects which pervade the rest of the economic system. There is indeed some evidence within the military sector on this point. It is frequently asserted that the Soviet Union devotes more resources than the United States to military research and development, with many more scientists and engineers engaged in military work. If this is the case, then the Soviet resources must be used in a relatively inefficient manner, for it is not disputed by the US Department of Defense that the United States continues to have a significant lead in most areas of military technology.

The book sets out some figures of Soviet output of various items of military hardware (table 2.1). These figures in general show high *levels* of output. They do not, in general, show sharply rising *trends*. Some of the figures do indeed show production rising—from 900 pieces of towed field artillery in 1976 to 1 300 pieces in 1980, for example. At least as many show declining rates. Under the constraint of the SALT I and SALT II agreements, the production of intercontinental ballistic missiles has come down

Table 2.1. US Department of Defense estimates of Soviet output of certain military items

Military item	1976	1977	1978	1979	1980
Ground forces materiel					
Tanks	2 500	2 500	2 500	3 000	3 000
Other armoured fighting vehicles	4 500	4 500	5 500	5 500	5 500
Towed field artillery	900	1 300	1 500	1 500	1 300
Self-propelled field artillery	900	950	650	250	150
Multiple rocket launchers	500	550	550	450	300
Self-propelled AA artillery	500	500	100	100	100
Towed AA artillery	500	250	100	-	-
Aircraft					
Bombers	25	30	30	30	30
Fighters/fighter-bombers	1 200	1 200	1 300	1 300	1 300
Transports	450	400	400	400	350
Trainers	50	50	50	25	25
ASW	5	10	10	10	10
Helicopters	1 400	900	600	700	750
Missiles					
ICBMs	300	300	200	200	200
IRBMs	50	100	100	100	100
SRBMs	100	200	250	300	300
SLCMs	600	600	600	700	700
SLBMs	150	175	225	175	175
ASMs	1 500	1 500	1 500	1 500	1 500
SAMs	40 000	50 000	50 000	50 000	50 000
Naval ships					
Submarines	10	13	12	12	11
Major combatants	12	12	12	11	11
Minor combatants	58	56	52	48	52
Auxiliaries	4	6	4	7	5

Source: Soviet Military Power, US Department of Defense, 1981.

from 300 to 200 between 1976 and 1980. For most of the other series the trend in simple numbers (which, because of the process of product improvement, is of course an inadequate measure by itself) is flat.

This impression of roughly constant output in numbers is reinforced by examining some of the figures for periods earlier than 1976, which are not given in this book. It appears that as far back as 1966 the Soviet Union was producing 3 500 tanks a year—a figure which rose to around 4 500 in 1970, and has since fallen back to the current rate of around 3 000 [2]. In shipbuilding, if we take the production of major surface combatants—cruisers, destroyers and frigates with a displacement larger than 1 000 tons—we find the peak in numbers was in 1953, when some 40 such ships were delivered. These were, of course, much smaller, cheaper, less sophisticated vessels than the ones now being built. Soviet output of these much larger, more complex ships is now running at a fairly steady rate of about five a year, with an output of about six a year of a class of corvettes, of just under 1 000 tons, called Grisha; this makes up the figure of 11 for major combatants in table 2.1 [3].

In the Soviet Union, therefore, as in Western countries, the increase in output of military hardware is not properly measured by crude numbers of weapons; it is the process of product improvement which is all-important, as new, more sophisticated models replace old ones.

The Soviet Union has traditionally relied on large quantities of simple, durable and relatively cheap weapons well-suited for mass production. The continuity of political and military leadership facilitates the long-term planning of research and production. This means that arms production in the Soviet Union follows a model-by-model type of development. Proven weapon systems are further developed and refined into new and more sophisticated versions. By using this method, existing production lines are easily converted to production of the new model and steady production can be maintained. So the more capable weapons are produced at much the same rate as their technologically inferior predecessors.

The burden that this massive military programme imposes on the Soviet economy must be the more noticeable now that the Soviet economic growth rate has slowed down. The Soviet gross national product (GNP) was rising at an average annual rate of 6 per cent during the 1950s; this declined to 5.2 per cent in the 1960s, to 3.8 per cent in 1971–75, and to 3.1 per cent in 1976–79. Many forecasters expect a further slow-down. The Soviet economy is plagued by lagging productivity, labour shortages and food shortages. The East European economies in general have incurred very substantial debts to Western bankers, and it is generally believed (rightly or wrongly) that the Soviet Union could not allow them to default. The support of the regimes in Cuba, Viet Nam and, more recently, in Poland is proving increasingly costly. In a planned economy, with labour shortages, the opportunity cost of military expenditure—that is, the cost measured in terms of civil production forgone—is more immediately apparent than it is in Western economies where at present there is substantial unemployment and substantial spare capacity. It stands to reason that those persons and institutions which are concerned with the performance of the civil economy must covet some of the resources devoted to the military sector.

Certain comparisons

The book on Soviet military power provides virtually no comparison with the military capabilities of any other country; yet power is, of course, essentially a relative concept. Comparisons simply between the Soviet Union and the United States are too limited. As a recent US Senate Committee report comments:

> ... The Soviets may see themselves as surrounded by hostile forces with no strong allies to assist them. Of the sixteen nations with the largest defense budgets as of 1978, seven, including the United States, are members of NATO, one (Japan) has a bilateral defense treaty with the United States, and three (China, Saudi Arabia and Israel) are strongly anti-Soviet or pro-Western in orientation. Only three of these countries (USSR, East Germany, and Poland) are members of the Warsaw Pact, another (India) is pro-Soviet in orientation ... Soviet fears of the People's Republic of China (PRC) have grown in the last three years as the PRC improved her relations with Japan and the United States. These developments are likely to be seen as highly unfavourable to the Soviets. Moreover, the Soviets have failed to improve their cool relations with Japan which has been a major foreign policy setback for them. [4a]

Comparisons, in short, must take account of the Soviet Union's long border with a country which it considers hostile—China; they must also take account of the extent to which the military expenditure of the NATO countries in Europe exceeds that of the East European members of the Warsaw Treaty Organization (other than the Soviet Union). It further appears that the military capabilities of Spain will soon be added to the NATO total.

In sum, the USSR and the other WTO countries maintain an output of larger quantities of conventional weapons than the United States and NATO. These weapons are much more sophisticated than a decade ago; nonetheless the technological lag is still there. Economic growth rates in the Soviet Union and in WTO countries in general have slowed down considerably: the economic cost of military output, in terms of civil output forgone, is likely to become increasingly disturbing.

Power projection

During the last two decades the Soviet Union has clearly set itself to become, like the United States, a true world-wide power. The construction of a navy with ocean-going capacity has been a central part of this programme. In the early post-war years, the Soviet Navy was simply a coastal force consisting of small ships such as fast patrol boats and corvettes. The Soviet Navy is now second only to that of the United States. In recent years, the strength and range of the surface fleet have been greatly increased by the entering into service of two Kiev-class aircraft carriers, the 40 000-ton Berezina fleet replenishment ship, one Kirov-class nuclear-powered missile cruiser, the 13 000-ton *Ivan Rogov* amphibious landing ship and Sovremennyj- and Udaloy-class destroyers. These lead ships and those that will follow give the Soviet Union a much greater peace-time 'power projection' capacity than it had before. However, this capacity is still much inferior to that of the United States. For instance, while the total aircraft carrier force of the Soviet Union consists of two 37 000-ton ships carrying 12–14 Yak-36 carrier-based fighters, the United States has some 14 aircraft carriers with an average displacement of approximately 70 000 tons, each of which takes 70–90 naval aircraft. The Naval Air Force of the Soviet Union consists of approximately 755 aircraft while the Naval Air Force of the United States consists of some 1 450 aircraft. The amphibious assault and tank-landing capacity of the Soviet Navy is very limited compared to that of the United States, and the United States is clearly better placed in naval logistics. The USSR has increased its access to certain facilities—as in Ethiopia and South Yemen—which it did not have before; however, the United States is clearly in a better position as regards the total number of bases, and obviously in the number of ice-free ports. The Senate Committee on Armed Services concluded, in a report published in 1981:

At the present time, the United States has substantial advantages over the Soviet Union in traditional power projection forces. The United States is far more capable of inserting and sustaining a military force in distant areas. While the Soviets have large airborne forces and a militarily more capable merchant marine—especially in terms of its coordination with naval forces—the majority of Soviet forces suitable for power projection are embryonic compared to US forces. The US advantages in sea-based tactical air, amphibious forces and shipping, airlift and in-flight refuelable aircraft are substantial. However, some of these advantages are offset, at least in part, by the greater proximity of the Soviet Union to key world trouble spots—the Persian Gulf, Middle East, and Korea. [4b]

Research and development

The Soviet Union has in the past decade been attempting, with its very substantial research and development programme, to reduce the United States' lead in military technology. Some results of this can be seen, for instance, in the improved capabilities of Soviet tactical aircraft. Traditionally, these consisted of relatively simple short-range interceptors such as the MiG-21, primarily intended for defence purposes. The new

MiG-23/27s, MiG-25s and SU-24s all have higher speeds and payloads, more sophisticated electronics and longer range. However, they still lag technologically behind the US aircraft.

Indeed the US Department of Defense, in its assessment of the Soviet Union's relative position, which is published in *Soviet Military Power*, suggests only two or three technological areas in which the Soviet Union might have a lead. These are direct-energy weapons such as high-energy lasers, chemical warfare and some radio frequency devices. In the rest the United States' lead remains. For example the report states "The United States remains the world leader in the field of micro-electronics and computers . . . The average relative position or 'gap' is 3–5 years with a few outstanding developments following US technology by only 2 years, and some problem areas lagging by as much as 7 years" [1]. Systems using micro-electronics and computers are at the core of modern weapon technology, and will continue to be so in the foreseeable future.

III. The United States

Before President Reagan came to the White House, the decision had already been taken by the Carter Administration to increase US military spending substantially in real terms. President Carter, in his five-year defence projection presented in January 1981, put forward an initial 8 per cent rise in the military budget for the fiscal year 1981 with a 5 per cent growth path thereafter.[1] This was a dramatic change in trend from the course of military spending in the previous decade. From 1968 to 1975–76 US military spending was coming down from its Viet Nam peak. It then stayed roughly constant, in real terms, up to the turn of the decade.

The Reagan Administration, given that it had campaigned on the inadequacy of President Carter's defence plans, had little choice but to move the numbers up, and did this in its revised March budget. This budget put in a volume increase in total obligational authority of 12.4 per cent in 1981, and 14.6 in 1982, with a 7 per cent real growth rate thereafter. This figure of 7 per cent seems to have been put in on the basis that it was 2 per cent higher than President Carter's figure. It was not based on any costing of proposed programmes—these were to be filled in to take up the money later. As Mr Stockman, the Director of the Office of Management and Budget, acidly but indiscreetly remarked "The defense program . . . was just a bunch of numbers written on a piece of paper" [5].

[1] These are the figures for Total Obligational Authority—the amount the Administration is asking Congress to authorize it to spend. A good part of these authorizations will be for actual spending in subsequent fiscal years, so actual outlays lag behind the figures for obligational authority.

These very big increases in total obligational authority in 1981 and 1982 would, if realized, produce large increases in actual outlay in 1983–85 (table 2.2). The new budget proposed by the Administration for the fiscal year 1983 requests an increase in total obligational authority of 13.2 per cent, after adjustment for inflation. The new five-year projection envisages an average increase in actual outlays, in real terms, of over 8 per cent annually from now to 1987. It seems quite likely that this year there will be a stronger Congressional opposition to the Administration's proposals. However, unless there is a very radical change in policy, there is no doubt that big increases are in process for US military spending. The beginning of the new trend is already there. NATO estimates of US actual military outlay show a 3.7 per cent volume rise in 1980, and a 5.9 per cent preliminary estimate for 1981.

Table 2.2. The United States military budget: five-year projections

Budget	1981	1982	1983	1984	1985	1986	1987
President Carter's January 1981 budget							
Total obligational authority in current dollars (billion)	171	196	224	253	284	318	–
Real growth (percentage)	*7.8*	*5.3*	*5.0*	*5.0*	*5.0*	*5.0*	–
President Reagan's March 1981 budget							
Total obligational authority in current dollars (billion)	178	222	255	289	326	367	–
Real growth (percentage)	*12.4*	*14.6*	*7.3*	*7.0*	*7.0*	*7.0*	–
President Reagan's January 1982 budget							
Total obligational authority in current dollars (billion)	–	214	258	285	332	368	401
Real growth (percentage)	–	*12.7*	*13.2*	*4.6*	*10.4*	*5.4*	*3.8*
Estimated real growth in outlays (percentage)	–	*7.7*	*10.5*	*8.0*	*9.6*	*8.0*	*4.6*

Note: These figures do not include the nuclear part of nuclear weapons, estimated at $4.5 billion in 1982, or military aid, estimated at $1 billion in 1982.

Source: Department of Defense Authorization for Appropriations for Fiscal Year 1982; Defense Daily, 9 February 1982.

Three issues about this substantial rearmament programme are considered here. The first concerns the reasons for its adoption. Secondly, there is the question of the form which the programme takes. Thirdly, there is a discussion of the economic consequences.

Reasons for adoption

The decision to change the trend in US military spending was not a reaction to any assumed change in trend in the Soviet Union. The CIA's estimates of Soviet military spending have for a very long time shown a steady and relatively unvarying upward trend. On the basis of the CIA figures, US Secretaries of Defense have indeed argued that the Soviet Union was outspending the United States—an argument summed up in the phrase 'When we build, they build; when we stop building, they build'. However, there is nothing new about this argument—and in fact in recent years it has come to be more widely accepted that the CIA's dollar estimates of Soviet military spending produce an overstated figure. Further, even on the CIA figures, total NATO military expenditure exceeds that of the WTO—and that is without including China's military spending.

The change in trend was rather the consequence of a change in public attitudes in the USA towards foreign policy and towards defence. This change has been summarized as follows:

By the end of 1980, a series of events had shaken us out of our soul-searching and into a new, outward-looking state of mind. The public had grown sceptical of detente and distressed by American impotence in countering the December 1979 Soviet invasion of Afghanistan. It felt bullied by OPEC, humiliated by the Ayatollah Khomeini, tricked by Castro, out-traded by Japan and out-gunned by the Russians. By the time of the 1980 presidential elections, fearing that America was losing control over its foreign affairs, voters were more than ever ready to exorcise the ghost of Vietnam and replace it with a new posture of American assertiveness.

Americans have become surprisingly explicit about how the United States should seek to regain control of its destiny, and in the context of the disquieting realities of the 1980s, these ideas created a new, different and complex foreign policy mandate for the Reagan presidency. The national pride has been deeply wounded: Americans are fiercely determined to restore our honor and respect abroad. This outlook makes it easy for the Reagan Administration to win support for bold assertive initiatives, but much more difficult to shape a consensus behind policies that involve compromise, subtlety, patience, restrained gestures, prior consultation with allies, and the deft geopolitical manoeuvring that is required when one is no longer the world's preeminent locus of military and economic power. [6]

This change in public attitude is statistically recorded in the Gallup poll which is regularly conducted on the public's views about defence. The response to a question on defence spending, expressed as a percentage of total replies, is shown below:

	1969	1976	1980
Too much	52	36	14
About right	31	32	24
Too little	8	22	49

The nature of the programme

To justify a drastic expansion of military expenditure (which was already scheduled to rise fast), the new Administration might have unveiled a new strategy. It did not do so: rather it has gone for an across-the-board increase in the acquisition of new weapons. The new Secretary of Defense told the Senate Armed Services Committee: "The principal shortcoming of the defense budget we inherited is not so much that it omitted critical programs entirely in order to fully fund others but rather that it failed to provide full funding for many programs it conceded were necessary but felt unable to afford" [7].

The programme includes big increases in expenditure on strategic nuclear weapons, the build-up of a much bigger navy and increases in the firepower and mobility of the Army and the Marines. The strategic nuclear weapon programme is discussed separately in chapter 4. The other two areas of increased expenditure have as their main purpose an increase in the ability of the United States to project its power in parts of the world which are distant from the US continent. The Secretary of Defense stated that the United States must be able to defend itself in "wars of any size and shape and in any region where we have vital interests . . . . Our global interest and commitments dictate that our armed forces acquire greater range, mobility and survivability . . . That means naval power able to command the sea lanes vital to us and our allies. It means developing urgently a better ability to respond to crises far from our shores and to stay there as long as necessary" [8].

For the Navy, the aim is to reach a 600-strong fleet by 1987: that means procuring some 143 combat ships. The long-term plan includes two new nuclear-powered aircraft carriers as well as the reactivation of four battleships and two aircraft carriers, 14 new attack submarines of the Los Angeles-class and some 1 900 aircraft, mainly F-18 fighters. For the fiscal year 1982 the main items include funding for one new aircraft carrier, reactivation of the World War II battleships *Iowa* and *New Jersey*, Aegis-class missile cruisers, and FFG 7-class frigates. Thirty F-14 and 63 F-18 carrier-based fighter aircraft will be procured during the fiscal year.

The additional funds (over and above the Carter budget) requested for the Army and the Marines are mainly intended for the Rapid Deployment Force, the emergency task force for rapid military operations abroad, primarily in the Middle East and the Indian Ocean. The fiscal year 1982 programme includes M-1 and M-60 tanks, M-2 infantry fighting vehicles, divisional air defence systems (DIVAD), attack helicopters, transport aircraft and AV-8B Harrier short take-off and landing (STOL) fighters.

One general consequence of the nature of the programme is an increase in the share of procurement in total US defence expenditure—from 24 per cent in 1980 to 30 per cent in 1982.

The major share of the new military orders will naturally go to the established defence contractors. McDonnell-Douglas is involved in three major aircraft programmes, namely the F-15 Eagle, the F-18 Hornet (in partnership with Northrop) and the AV-8B Harrier (in partnership with British

Aerospace). General Dynamics is producing the F-16 fighter for which the US Air Force alone has an order for 1 388 aircraft; the company is also building the Ohio- and Los Angeles-class nuclear submarines, as well as manufacturing various ship-borne missile systems. Tenneco is responsible for the construction of nuclear aircraft carriers, Chrysler is the main contractor for the M-1 Abrams tank, and Raytheon and Hughes manufacture Maverick, Phoenix, Sparrow and Sidewinder airborne missile systems. They are also co-developing the new NATO medium-range missile called AMRAAM.

The economic consequences

There has been considerable debate in the United States about the economic consequences of this military spending programme, with some economists arguing that it will wreck the economy and others saying that it can be accommodated with no great difficulty. These are some of the points made in that debate:

1. First of all, there is the question of whether the programmes which have been launched will not demand even larger budgets than those now put forward. One virtually universal characteristic of weapon procurement programmes is that they overrun their initial estimates. The average cost overrun of major programmes—not including inflation and quantity changes—has been put at nearly 52 per cent; the chance of a major programme being completed within its initial cost estimate is about one in ten. There will almost certainly be strong pressure from the three services for higher budget allocations.

2. There is, of course, no dispute that this programme will raise the share of military expenditure in the national product. However, how big that rise will be depends crucially on the rate of growth of US GNP: and this, in turn, will depend very largely on the extent of the recovery (if any) in the United States' productivity trend. In recent years, productivity in the United States has hardly been rising at all. The Reagan Administration claims that its 'supply-side' policies will rejuvenate US productivity: many economists doubt it. If productivity recovers to a 3 per cent trend, military spending (in the present programme) will go up from 5.7 per cent of GNP in 1981 to 7.1 per cent in 1986. If there is no recovery in productivity, that 1986 figure becomes 8.1 per cent [9].

3. The critics who suggest damaging economic consequences from this military programme do so mainly on the basis of the general economic policies which, under the present Administration, seem likely to accompany it. The critics do not dispute that, with appropriate economic policies,

room could be made in the economy to allocate 7–8 per cent of the national product to military spending. It is still, after all, a lower percentage than the average for the 1950s, which was around 10 per cent. However, unless there is substantial spare capacity in the economy (a point discussed below) a relative increase in military demand for resources requires policies to produce a relative decrease in civil demand. The Reagan Administration, it is true, is proposing reductions in federal civil expenditure: however, it is also proposing reductions in tax rates.

4. The inflationary dangers from the proposed military programmes are twofold. There is, first, the 'bottleneck inflation' which comes from specific shortage of materials or skilled personnel needed for these weapons programmes and, secondly, there is also the risk of general excess demand inflation. The first of these is virtually certain, the second is more controversial.

In the period when weapon procurement in the United States stagnated, many sub-contractors who had previously been largely engaged in military work turned to civil production. As a result, when military orders increased, bottlenecks appeared as early as the autumn of 1980. A Congressional Committee received testimony at that time, that "from 1976 to 1980 the typical delivery span of aluminium forgings increased from 20 to 120 weeks . . . From 1977 to 1980 the delivery span for aircraft landing gear grew from 52 to 120 weeks . . . In spite of the recession and its attendant unemployment, there remains a shortage of skills needed by the defense industry. The shortage leads to competition for labor and upward pressure on costs" [10]. It is probable that, with the deepening recession since 1980, these delivery times will have shortened. Once the economy begins to recover, they could soon lengthen again.

Whether general excess demand inflation will follow from these programmes is more controversial. There is no consensus estimate in the United States of the extent of spare capacity in the economy now; nor, of course, is there a consensus view among economists about the determinants of inflation. Some will regard the federal budget deficit as the crucial figure in this regard; others will look rather at the figure for unemployment, as a general measure of the pressure of demand. It would not be sensible to attempt to make a five-year forecast of the course of unemployment, simply on the basis of the military expenditure programme.

5. Finally, there is the question of the effect of the increased demand for weaponry on US high-technology civilian industries, as materials, equipment and skilled personnel are moved from civilian to military pursuits [11]. This does seem likely to do some damage to the ability of the USA to compete in world markets, and unless the USA turns to more trade protection there will be a loss of share in the home market as well.

For the United States is the only Western industrial country which is rearming rapidly. In Western Europe, and more particularly in Japan, the demands of military high technology will not be bidding resources away from the civil sector. US high-technology firms which produce civil products will find it increasingly hard to hold on to their markets.

The future course of US military spending is much more likely to be determined by these economic factors, and by any consequent changes in public attitudes, than it is by any sophisticated analysis of the Soviet threat.

IV. The NATO targets

In May 1977 NATO countries collectively agreed to begin to move their military expenditure up to a 3 per cent real growth trend; this undertaking was repeated in May 1978, and again in May 1981, when the period was extended to 1987. When ministers agreed on these 3 per cent growth targets, they probably thought that there were clear and unambiguous figures for volume increases in military expenditure. This is, after all, a very common view among those unacquainted with the statistical complexities of such a calculation. One reason for expressing the NATO target in these terms was, no doubt, because ministers had been told that Soviet military expenditure had been rising in volume terms by 3 (or 4 or 5) per cent a year for a long period: so the best thing to do was for NATO countries to do the same.

In fact, these figures are anything but clear and unambiguous, as subsequent events and arguments have shown. First of all, there are a number of alternative series for military expenditure—budget figures and outlay figures, figures including or excluding military aid, national figures and standardized NATO figures, and so on. Secondly, it was never clear what base year was to be used for these calculations—and, given that there are some erratic year-to-year movements in military spending, the choice of a base year can make quite a difference. Thirdly, there seems to have been no discussion of the price indices which should be used for the volume calculation. Some countries have a specific price index for the military sector, others do not. There are great problems in constructing a sensible price index for sectors where 'product improvement' is rapid—and the military sector is one such sector.

There is an interesting illustration of this problem in arguments in the UK about the proposed military budget for 1982/83. The Treasury has tentatively put in an 11.4 per cent money increase for 1982/83 over the revised figure for 1981/82, arguing that this allows for 8 per cent inflation and consequently permits a 3.4 per cent volume increase, which meets the 3 per cent target and leaves a margin. The service chiefs complain that the rate of inflation in the defence sector is at least 2 per cent higher than in the economy as a whole, with equipment costs going up by around 14 per cent a year and sometimes more. Then the question arises: how much of this 14 per cent is really a price increase, and how much is the consequence of 'product improvement', and thus should be counted as a volume rise? Even within individual countries, there has been no agreement about the meaning of the target.

In discussing their country's compliance with the NATO target, ministers can pick and choose among a number of different possible military expenditure series and calculations. One of the curiosities of this situation is that, in the discussion of this question, not much use seems to be made of the NATO standardized figures for military expenditure. After all, these figures have been prepared with precisely this purpose in mind—to make comparative statements about NATO countries' military performance which are fair, because the figures are standardized.

Table 2.3 uses these NATO standardized figures to try to answer the question of whether or not NATO countries have accelerated the growth of their military spending since the 3 per cent volume target was adopted. NATO figures are all 'outlay' figures—that is, estimates of actual expenditure, not budget forecasts—and they are all on a calendar year basis. The table, to avoid the problem of erratic base years, uses the average of three

Table 2.3. NATO countries: estimated volume increases in military expenditure

	Per cent increases			
Country	'Pre-target': From 1972–74 average to 1976–78 average	'Post-target': From 1976–78 average to 1981	Latest year: From 1980 to 1981 (estimated)	Size of military spending in relation to USA (USA = 100)[a]
United States	−2.0	3.0	5.9	*100*
Canada	3.9	0.4	1.9	*3*
All NATO Europe	2.3	2.1	1.0	*74*
of which				
FR Germany	1.0	1.7	1.7	*20*
France	3.8	3.0	2.0	*18*
UK	0.3	2.3	−3.6	*16*
Italy	−0.4	4.1	0.5	*6*
Netherlands	3.4	1.3	0.6	*4*
Belgium	5.1	2.3	−0.3	*3*
Turkey	16.0	−1.4	21.1	*2*
Greece	14.4	−2.5	4.1	*2*
Denmark	3.2	1.3	0.8	*1*
Norway	4.1	2.4	1.0	*1*
Portugal	−13.5	2.5	2.4	*1*
Luxembourg	5.8	7.0	4.5	neg

Source: SIPRI Yearbook 1982, table 5B.2.
[a] Based on 1980 military spending figures, at 1979 prices and exchange-rates.

years for this purpose. For producing volume series, consumer price indices are used throughout: some countries have specific indices for the military and others do not. However, with alternative price indices it is most unlikely that the general conclusions which follow would be changed.

The conclusions are fairly straightforward. The United States has turned round the volume of its military spending. It had been falling back from the high Viet Nam peak until 1976; since then it has been rising on an accelerating trend. The 1981 estimated increase is nearly 6 per cent, and present plans call for about an 8 per cent volume increase from now on. How far such a massive increase will in fact be realized is obviously a matter for debate: it is discussed in the US section. However, unless there is a very radical change in policy, there is no doubt that the United States will exceed the 3 per cent target by a wide margin.

The story for other NATO countries is a very different one. For NATO countries in Europe in total, and for Canada, there has been a *deceleration*, not an *acceleration*, in the volume growth of military spending since the target was announced (see table 2.3). In the four pre-target years, military spending in NATO Europe was rising at 2.3 per cent a year; since the target was announced, the figure has come down to 2.1 per cent, and preliminary estimates for 1981 show only a 1 per cent rise. In Canada the change is even more marked; since the target announcement, there has been hardly any rise at all in military spending in real terms. The United States has begun a formidable programme of rearmament. NATO Europe and Canada have not.

There are a number of reasons for these very different patterns of behaviour. In the United States a great many people have felt that the United States' status as a great world power was being challenged—by humiliation in Iran, and by a much increased Soviet threat. Politicians in European countries, on the other hand, did not in general see any radical change in the position in Europe: there did not seem any particular reason to think that the Soviet threat in Europe had suddenly become more acute. Indeed, their reaction to the United States' rearmament programme—an implicit and not, of course, explicit reaction—may well have been that, with the United States accelerating its military spending so much, there was really no need for them to do the same.

West European countries were much more preoccupied with their economic problems—particularly with the problem of inflation, which (rightly or wrongly) was widely attributed to their budget deficits. (Even with the slowing down in the rate of increase in military spending, it has been increasing as a percentage of the 'NATO European' national product —from 3.6 per cent in 1979 to an estimated 3.8 per cent in 1981.)

For those concerned to see a reduction rather than an increase in world military spending, it is a source of some relief that, up to now, the European

NATO countries have not done what they said they would do. There are dangers, however. The US pressure on West European countries to 'carry more of the burden' will undoubtedly intensify: there will be more Senators asking the question 'Why should we defend the Europeans, if they are not willing to defend themselves?' Secondly, we may well see in Western Europe a swing back to Keynesian reflationary policies, and a swing away from preoccupation with budget deficits: we can then expect defence ministers to put forward the argument that rearmament will create jobs (an argument which is already widely used in the discussion of individual weapon programmes).

V. Western Europe

There is a common theme in the story of military expenditure in Western Europe in 1981: a conflict between the rising costs of new weapon systems on the one hand, and a desire to reduce budget deficits on the other. In a number of countries, weapon procurement costs have outrun their budgets, not only for the usual reason of cost overruns, but also for other reasons connected with the general economic recession in the West. The firms producing both civil and military goods have found that the influx of orders for their civil production has been much reduced, so they have completed their military orders on time, or indeed early, and have expected payment. So military expenditures have tended to exceed the budgeted figure. This has come at a time when in a number of countries the reduction of budget deficits has become central to the government's anti-inflationary strategy. Defence ministers and finance ministers have thus come into sharp conflict: in a number of West European countries there have been defence reviews of one kind or another, and weapon programmes have been reduced in an attempt to keep the military budget down.

The United Kingdom

The NATO standardized figures for the UK's military spending show a rather strange year-by-year pattern: a big increase in 1979 (of 4.5 per cent in real terms) followed by an even bigger rise in 1980 (of 8 per cent) and then, on provisional figures, a fall in 1981. A better impression of what is happening is given by grouping the last two years together. After a long period, from 1972 to 1978, in which military spending in the UK was running virtually flat, it is now on a rising trend of the order of 3 per cent a year in real terms.

However, in spite of an explicit decision to change the trend in military spending, the UK has also encountered a sharp conflict between rising weapon costs and budget constraints. In the fiscal year 1980/81, military spending exceeded the initial budget by some £500 million, and there had to be two supplementary estimates. In the fiscal year 1981/82, spending in excess of the original budget of £12.3 billion may be £700 million. In April 1981, the Defence White Paper gave notice of a thorough review of defence spending, one more such review in the very long series of defence reviews in the UK. The main points of the review which emerged in June 1981 can be summarized as follows:

1. The 'independent nuclear deterrent' was sacrosanct.
2. The main cuts were to be made in the Navy's surface fleet.
3. A new general principle was pronounced, that the UK was spending too much money on weapon platforms and too little money on the weapons to go on these platforms.

Although (as ministers constantly point out) the independent nuclear deterrent forms a very small part of the UK's military spending, it is the item in the military budget which has been most discussed. First of all, there has been a very expensive programme of upgrading the warheads on the Polaris missiles; this programme has gone under the label 'Chevaline'. Its total cost was put at about £1 billion in January 1980, when it was thought to be almost completed; since then, significant further expenditure must have been incurred. The main purpose of the programme was to ensure that Polaris missiles would be able to penetrate any further upgrading of Moscow's ABM defences; Moscow is the only city in the Soviet Union which has any ballistic missile defence. The existence of the programme only became known when it had nearly been completed; there was thus no public discussion of its necessity. Apart from the whole question of the value of an independent nuclear deterrent, it is not clear why Moscow itself has to be the target, rather than other Soviet cities which do not have ABM systems. It has been suggested that Chevaline may have been undertaken simply to improve the UK's indigenous capability in warhead construction [12].

The Chevaline programme apparently ran into considerable trouble; in the early tests there were difficulties over the separation process when the warheads and decoys were detached; and a fresh series of tests was started early in 1982. This is not the end of expenditure on the Polaris system. Work has begun, also early in 1982, on replacing the motors in the nuclear missiles. This programme will cost several hundred million pounds spread over a number of years.

The government proposes to replace the Polaris missile system with the US Trident missile system, on a new fleet of submarines. Here a problem has arisen because of the US Administration's decision to press ahead with the Trident II missile. The original British plan was to employ the Trident I missile; however, by the time British submarines are ready, the USA will probably be phasing this missile out, and there would be great difficulties in maintaining in the UK a missile which was no longer in operational use in the United States [13]. The decision to change to a system built on the Trident II missile would have a number of complications. First of all, there would be a significant loss, estimated at £50 million, in 'long-lead orders' which have already been given on the previously existing plans. Secondly, the Trident II is both fatter and longer than the Trident I missile, and the submarines built to carry it would have to be larger—probably at least 15 000–19 000 tons rather than the originally planned 10 000–12 000 tons. Further, both the Trident I and Trident II missiles are weapons of far greater sophistication and accuracy than is needed for a deterrent (rather than a counterforce) weapon. The total cost of replacing the Polaris system may be around £7 billion.

The June Defence Review introduced substantial cuts in the Navy's surface fleet; the Navy's main role is to be an anti-submarine one, and for this purpose there would be greater emphasis on nuclear-powered attack submarines and Nimrod maritime patrol aircraft. Whereas in 1981 there were some 63 ships of frigate size and above in the UK's surface fleet, by 1985–86 that number is to be reduced to 44 [14]. The number of nuclear-powered attack submarines, on the other hand, is to rise from 12 to 17. The Nimrod aircraft would be armed with Sting Ray lightweight anti-submarine torpedoes; and in September it was decided that the contract for a new heavyweight torpedo for the Navy's attack submarines would be awarded to Marconi Space and Defence Systems. This will replace the newly introduced Tigerfish torpedo which is not considered capable of destroying the new deep-diving Alpha-class Soviet submarines with two-layer titanium hulls.

Whereas in FR Germany there has been a great deal of discussion about the Tornado programme, in the UK there has been much less attention paid to it, although it is much more expensive than the cost of replacing the Polaris system. The total cost for the UK of the Tornado programme has now been put at £11 250 million; expenditure on that programme is now reaching a peak in the UK as in FR Germany. Consequently, in agreement with FR Germany, the UK has reduced the peak annual delivery rate of these aircraft from just over 60 to 44, again as part of an attempt to keep military expenditure at or near the budgetary figure. However, the total Tornado programme, which is due to be completed by 1988, has so far remained the same: a total of 385 aircraft for the UK, of which 220 would be the interdiction-strike version (IDS) and 165 the air-defence variant (ADV).

FR Germany

FR Germany has been much more resistant than the United Kingdom to pressures to accelerate its military spending. The rate of increase in recent years has only been of the order of 1.5 per cent a year in real terms: and it is doubtful whether the rise in 1982 will be any greater than that.

In FR Germany discussion of the military budget has been much preoccupied with the costs of the Tornado. This is not surprising, given that the original budget estimate for the Tornado programme in 1981 was DM 1 750 million and a series of upward revisions has brought the figure up to DM 3 065 million [15]. The Tornado programme is proving an immensely expensive one and is leading to cutbacks in weapons procurement elsewhere.

The story of the Tornado—originally referred to as the multi-role combat aircraft (MRCA)—goes back as far as April 1965, when the United Kingdom cancelled its own programme for the TSR-2. The British government then turned to examine the possibility of some collaborative arrangement with France; these negotiations broke down in 1967. The UK then turned to other European countries which had aircraft industries in need of work, and which also had a requirement to replace the F-104 Starfighter. Eventually in 1970 FR Germany, Italy and the UK agreed on a joint programme. The main attraction of the programme was that it would help all three countries to keep an aerospace industry in business, and it would help to maintain some European independence from the USA in the production of military aircraft.

The original unit fly-away price was put in 1970 as DM 15 million: the fly-away price includes costs of production, acceptance flights, and other recurring costs. The original unit system price was DM 28 million; this includes spares, ground and training equipment, armaments transport and packing, and so on. By the end of 1980 these two figures had risen to DM 40 million and DM 70 million, respectively. Part of these rises was of course caused by general inflation; however, when the figures are corrected for the general rate of inflation in FR Germany over the decade, this still leaves an overrun in the unit cost, in real terms, of 50–60 per cent. Furthermore, it seems that the cost of the plane is still rising significantly faster than the general rate of inflation in FR Germany. Various sources suggest that by the end of this year the unit system price may be nearer DM 100 million [16, 17].

FR Germany, like the UK, is faced with a very large total bill for its Tornado programme, although it plans to procure a rather smaller number than the United Kingdom: the West German order is for 324 of the interdiction-strike version. It has agreed with the United Kingdom and Italy to cut back the peak rate of production. Even so, it has had to cut back on the procurement plans of a number of other weapon systems. Ironically one of the programmes that has been cancelled in favour of continued Tornado production is the programme for 200 Roland air-defence missile systems. The Roland units were supposed to protect Tornado and NATO AWACS airfields in FR Germany against low-level attacks. Also cancelled are 2 000 MILAN anti-tank missiles from the Franco–German Euromissile consortium, and research and development on the TKF-90 tactical combat aircraft programme. The planned collaboration with France on a main battle tank for the 1990s to succeed the Leopard and AMX-30, and on the PAH-2 anti-tank helicopter, now seems, if not dead, at least highly uncertain. Among major weapon systems that have been postponed are two additional Type 122 frigates, and Gepard anti-aircraft vehicles.

France

In France, the trend in military spending has been for a rise, in real terms, of over 3 per cent a year over the whole of the past decade. It is a trend which seems likely to continue. Military spending in France appears to be a relatively uncontroversial issue: it was not an issue in the presidential campaign. The French government can take actions in the military field which other European governments would find extremely difficult: thus the new government has announced that France will continue testing neutron weapons and would not rule out their deployment with its national forces. In many other countries, an announcement of this kind would be met with a storm of protest.

The new Administration has made no significant change in the military policy of its predecessors. It is a policy of independence within the Atlantic Alliance; the Defence Minister in the new Administration, M. Charles Hernu, said recently "We must keep our freedom to make decisions, without automatically becoming involved in a conflict against our will" [18].

Thus the policy is not simply to maintain the nuclear deterrent, but to develop it: the new Administration has agreed in principle to go ahead with the construction of a seventh nuclear missile submarine; and the nuclear test programme at Mururoa is to continue. Further, French military policy is not exclusively concerned with Europe. There are agreements with certain African countries which mean that France considers it should have the means for external intervention and should equip itself with this in mind. Thus there is a French Rapid Deployment Force, which has now been increased to some 20 000 men, and has the capability of intervening in the former French colonies in Africa.

Changes are under way in the ownership structure of the French arms industry, as a substantial part of this industry becomes nationalized: for instance, Dassault-Breguet (Mirage aircraft), Matra (missiles) and Thomson-Brandt (defence electronics). It remains to be seen whether this will make any significant difference to the behaviour of these companies.

Three smaller countries

On NATO provisional estimates of military expenditure in 1981, Belgium, Denmark and the Netherlands all had roughly zero growth in military spending (in real terms). In all three countries, the need to hold back public expenditure overrode the commitment to the NATO 3 per cent target. For 1982, the Netherlands government has put a 3 per cent real increase in military spending into its estimates. In Belgium, on the other hand, the 1982 budget would seem to imply a decrease (in real terms). Procurement in particular seems likely to be held back. No funding, for example, is provided for the replacement of the 80 Mirage 5s in the Belgian Air Force, before 1984. In Denmark, the minority government initially proposed a freeze on defence spending in real terms. It came under pressure to change that proposal from NATO in general and from the United States and Norway in particular. As a consequence an agreement was reached with the main opposition parties on a programme which would increase military spending in real terms by 1 per cent in 1982 and 0.5 per cent in both 1983 and 1984.

VI. Japan

Article 9 of the Japanese constitution is part of the necessary background to any discussion of Japanese military expenditure and policy. At the end of World War II, the United States imposed upon Japan a constitution which explicitly forbade the maintenance of military forces. Article 9 reads as follows:

> Aspiring sincerely to an international peace based on justice and order, the Japanese people forever renounce war as a sovereign right of the nation and the threat or use of force as a means of settling international disputes.
>
> In order to accomplish the aim of the preceding paragraph, land, sea and air forces, as well as other war potential, will never be maintained. The right of belligerency of the state will not be recognised. [19a]

It has obviously been something of a problem to reconcile the gradual reconstruction of Japanese armed forces with this article. One consequence has been a succession of semantic changes. The armed forces were initially called a national police reserve; then they became known as the 'Safety Forces', under a 'National Safety Agency'; finally they have become the Ground Self-Defense Force (GSDF), the Maritime Self-Defense Force

(MSDF), and the Air Self-Defense Force (ASDF), under the Japanese Defense Agency.

However, article 9 of the constitution is still important in the public mind. It might have been expected that, since it was imposed on Japan by an occupying power, there would be strong public opposition to it. In fact, that does not appear to be the case. In a public opinion survey early last year, in answer to the question 'Is it desirable or not to amend Article 9 of the Constitution so that Japan can possess fullfledged armed forces?', 71 per cent replied that it was not desirable. The majority took the view that the Self-Defense Forces were not against the constitution [19b]. However, although most people were in favour of the existence of the Self-Defense Forces, it is only in recent years that their main function was seen as the maintenance of security; in the early 1970s people were more concerned that the Self-Defense Forces should engage in disaster relief operations (figure 2.1). Nor is there any significant public pressure for more military spending: in surveys in the spring of 1981, the majority opinion was that the Self-Defense Forces 'ought to stay at the present level of strength'.

The pressure on Japan to increase its military spending, therefore, does not come from public opinion. It comes mainly from the United States, which has of course long since abandoned its objective of demilitarizing Japan. The United States Administration clearly feels that Japan is a free

Figure 2.1. Japanese public opinion survey: on what should the Japanese Self-Defense Forces concentrate?

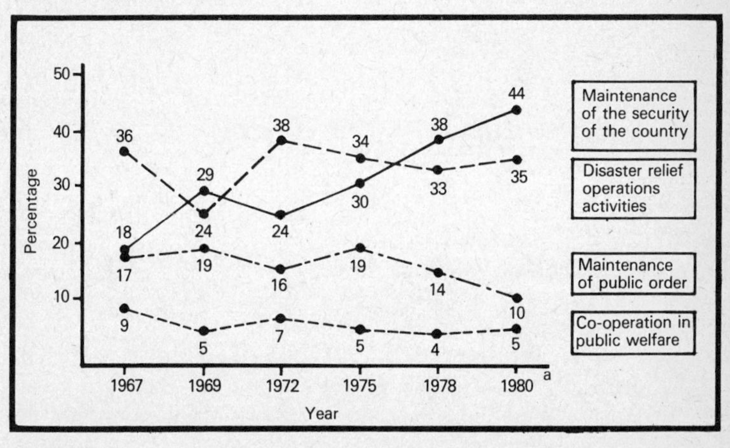

[a] The 1980 question was not precisely the same as that of previous years.

Source: Survey by Japanese Prime Minister's Office, quoted in *Defense of Japan 1981*, Japanese Defense Agency.

rider in military matters. Further, in the view of the United States, Japan's low percentage of resources devoted to military uses has helped it to develop a highly competitive civil industry. So Japanese products are reducing the US share of the market both overseas and in the United States itself as well. Thus there is a link between Japan's low military posture and its large trade surplus with the United States.

Under the previous US Administration, the pressure on Japan was simply, in a general way, to spend more on defence. Under the new Administration, there is a more specific suggestion: that Japan, in addition to providing for the self-defence of its own islands, should also defend the airspace and sea lanes up to 1 000 miles from its shoreline [20]. This would obviously require new weapons, such as attack submarines and new aircraft. The suggestion has been met with a cool response in Japan. There may also be some pressure from the large corporations, which are showing some interest in moving into military production in a more substantial way. Thus the chairman of Mitsubishi Corporation, Mr Bunichiro Tanabe, is recently on record as saying, "It is about time the Government lifted the ban on exports of arms to foreign countries" [21].

The long-term trend in Japanese military expenditure, from 1971 to 1979, has been for a real rise which averaged rather more than 4 per cent a year. There was a check in 1980, and then the real rise was resumed in 1981. The budget for 1982 is for a rise of 7.5 per cent in military spending—at a time of zero growth in other categories of government expenditure. This figure does not include any increase in the pay of the armed forces: there will, therefore, be a further real rise this year.

However, the Japanese government will keep to its unwritten rule that military expenditure should not exceed 1 per cent of GNP. Even so, its military spending places it fifth among Western industrial countries.

VII. China

For some time, China has been following a more open policy on publishing material about its military expenditure. The 1981 figure shows an interesting movement. Whereas most other major countries were either increasing, or at least maintaining, their military spending in real terms, in China the military budget for 1981 was cut substantially. The reduction from the 1980 figure was no less than 13 per cent (in current yuan), and the military sector's share of the total national budget fell significantly.

China's assessment of the threat from the Soviet Union is clearly very different from that of NATO in general, or the United States in particular. It has obviously come to the conclusion that there is currently not much risk of a Soviet attack. It is now giving top priority to the improvement of its civil economy. It may also have taken the view that, since a number of other countries are engaged in increasing their military strength *vis-à-vis* the Soviet Union, there is less need, not more, for China to do the same.

Military spending has therefore clearly been given a lower priority in China at the moment. Indeed there are reports that some of China's large weapon production plants (which were originally constructed on the massive scale of Soviet plants of the same kind) are being partially converted to the production of consumer goods.

One consequence of this low priority to military spending in China at the moment is that the US and European firms which had hoped to sell large quantities of military hardware to China are likely to be disappointed. The purchase of foreign military equipment is not high on China's priority list for the expenditure of scarce foreign currency.

The figures in table 2.4 include procurement. The responsibility for the production of weapons rests with the production ministry concerned; most production ministries are responsible for both civil and military production. The Defence Ministry then purchases the weapons from the production ministries. Although procurement is included in the figures, it is probable that the bulk of military research and development expenditure is not included. However, although this exclusion will affect the estimates of the level of military spending, it is unlikely that it alters the trend.

Table 2.4. China's budget figures for military expenditure

Year	Billion yuan
1977	14.9
1978	16.8
1979	20.2[a]
1980	19.4
1981	16.9

[a] The budget figure. Because of the war with Viet Nam, actual expenditure probably exceeded this figure by about 2 billion yuan.

The number in the armed forces is of the order of 4 million men. The higher estimates that have been given—of 4.75 million—probably include the railway, construction and engineering regiments. These, although they still exist as units, have been transferred to civilian control, and are primarily engaged in civil work: for example, the construction of the Peking underground system.

The conversion of the yuan figure into dollars presents the usual problem: what exactly is the meaning of the figure when it has been converted? Indeed, there is little point in attempting a conversion, except in order to provide some kind of estimate for the world total. At the official exchange-rate, China's military expenditure in 1981 (making some allowance for research and development expenditure) would only be of the order of $10.5 billion.

However, the use of the official exchange-rate obviously makes little sense. The conscripts in China's army are paid very little. Conscription is for three years in the army, four years in the air force and five years in the navy. Conscripts have full provision of food, clothing and shelter, and in addition receive 7 yuan a month in the first year, 8 yuan in the second, and 9 yuan in the third. Total expenditure on military personnel in 1981, including food, clothing and shelter, was probably of the order of 5 billion yuan: that is, about 1 250 yuan a year for each member of the armed forces. The comparable figure for the United States is about $16 000 a year. That gives an exchange-rate for military personnel of $13 to the yuan. The cost of a military sector of 4 million persons in the United States would be of the order of $65 billion.

This is, of course, not a sensible figure. Where conscripts cost so little, the military authorities are of course lavish in their use of manpower. The search for a 'correct' dollar figure for the military expenditure of countries such as China, whose military and economic system is wholly different from that of the United States, is a search for a mirage. The only reason for giving a figure at all in appendix table 2B.1 is to provide some kind of estimate of the world total of resources which the military sector uses.

VIII. Some notes on other regions

The Middle East, North and East Africa

In both Egypt and Israel military expenditure (at constant prices) seems to have been coming down, for some years, from the peak period of 1973–77. However, the SIPRI figures of military expenditure include military aid as part of the spending of donor countries, not recipient countries; and now Egypt as well as Israel is receiving US military aid in substantial quantities. In the US fiscal year 1982 (ending on 30 September 1982) Israel will receive $1.4 billion in military aid; Egypt will receive $900 million. This figure for Egypt does not include some $500 million which will be spent on modernizing an air base and supply depot at Ras Banas on the Red Sea [22].

There are no reliable recent figures for the military spending of Iraq and Iran, still engaged in a desultory war. Other Arab states have been giving substantial assistance to Iraq: in April last year Kuwait granted a $2 billion interest-free loan, and there have also been loans from Saudi Arabia and the United Arab Emirates [23].

Military spending in the Middle East is now dominated by the very large figures for Saudi Arabia, which, as a rich country, does not receive military aid but buys its military equipment. However, military spending is also rising fast in some of the smaller states, such as Oman, where it is estimated to have doubled between 1979 and 1981. The flow of military aid from the two great powers to the countries around the Persian Gulf—and to North African countries—is increasing. There are reports of substantial Soviet arms caches in Libya, for example [24]. The United States is using military aid to win access to military facilities near the Persian Gulf, and there is now a formidable programme of US base construction. Oman has indicated that it expects the USA to spend some $1–1.5 billion on military facilities over the next 10 years [25]. Somalia has agreed to provide increased access to its air and port facilities, in exchange for aid. There has also been increased military aid to the Sudan and to Tunisia, and Kenya has agreed to allow US use of Kenyan facilities, notably the port of Mombasa and the airfields of Embakasi and Nanyuki.

South Africa

In South Africa, military expenditure is budgeted to rise sharply. There had been a previous spurt in South Africa's military build-up between 1974 and 1977, set off by Portugal's withdrawal from Angola and Mozambique. The figure then came down temporarily in 1978, partly because of the arms embargo imposed in 1977. Now, in the fiscal year 1981/82 (which runs from 1 April to 31 March), military spending is scheduled to increase by some 30 per cent in money terms, or 15 per cent in real terms (inflation is running currently at about 15 per cent).

So far as military procurement goes, the bulk of the appropriation goes to ARMSCOR, the state-owned Armaments Development Corporation. This Corporation controls directly or indirectly the production of most of South Africa's weapon requirements. South Africa produces (for some items under licence) the French Mirage aircraft, the Italian Aermacchi training aircraft, French-designed Panhard armoured cars, Israeli-designed missile boats, a derivative of the French Crotale surface-to-air missile, air-to-air missiles, artillery pieces, infantry weapons and a wide range of ammunition. In September 1981, the ARMSCOR chairman said that South Africa was now self-sufficient in ammunition; ARMSCOR subsidiaries and contractors manufacture 141 different kinds of ammunition for the army, air force and navy. ARMSCOR claims to be the West's tenth biggest arms producer. However, South Africa is still concerned with the clandestine acquisition of some items of military equipment. Thus it succeeded in acquiring from a Vermont-based production firm, Space Research Corporation, 50 000 155-mm howitzer shells

[26, 27]. This shell is said to be the most advanced product on the market, and to have applications in tactical nuclear warfare. The South African arms industry also appears to have links with Israel, Taiwan and some South American governments.

The army takes over half the defence budget; 80 per cent of all military personnel are in the army. In training exercises, emphasis is laid on counter-insurgency operations, commando strike techniques, and close air support of mobile ground operations. The focus of procurement is towards complete self-sufficiency in items that are being, and will be, needed in sustained, low-level operations.

Now that there is no longer co-operation with Western naval forces, the role of the South African Navy is changing. It no longer considers that it has an anti-submarine warfare (ASW) function to perform on behalf of other nations, and more emphasis is now put on local naval defence, with small strike craft, such as the Israeli-designed missile boats.

India and Pakistan

From 1972 to 1979, military expenditure did not rise much (in real terms) in either India or Pakistan. Since 1979, it has risen quite sharply in both countries. In India, military spending in 1981 is estimated to have been some 8 per cent higher (in real terms) than in 1979. In Pakistan, the rise over the same period was 20 per cent—but on a much smaller total. Further, these figures do not include military aid; and there may well have been arrangements outside the military budgets for the purchase of weapons from the United States and the Soviet Union. To that extent, the military expenditure figures underestimate the size of the increase in resources devoted to the military sector.

Any account of the course of military spending in these two countries is closely tied up with a discussion of the arms trade (chapter 3). In mid-1980 India completed a large arms deal with the Soviet Union, with a nominal interest charge and a 17-year period of repayment; there were also substantial purchases of Jaguar aircraft from the UK and arrangements for their local manufacture.

In Pakistan, the first offer of US aid, in 1980, was turned down because the sum was too small. About a year later, a new and much more substantial aid package was negotiated, for $3.2 billion for five years starting from October 1982; half the money was to be for military purchases. Implicit in the deal is the understanding that it is likely to be cancelled if Pakistan carries out a nuclear explosion. Pakistan is also probably receiving help from 'Islamic friends' to purchase F-16 aircraft before October 1982.

The United States, in giving this military aid, undoubtedly had the northern border of Pakistan in mind. The matter may well be viewed in a different light in India. At the time of writing, however, talks were in progress between Pakistan and India on a non-aggression treaty proposed by Islamabad.

Australia

After a long period of what might be called 'passivity' in military spending, it is now rising quite sharply in Australia (and also in New Zealand). Between 1974 and 1979 there was virtually no change—the rise in Australia was under 1 per cent a year in real terms. In the last two years, the increase has been (again in real terms) over 6 per cent a year.

This is a reaction to the general world situation, rather than to any perceived threat: reaction, that is, to events distant from Australia. Indeed one of the problems of Australian defence policy has been to decide for what range of contingencies to prepare. The policy has been to develop a core force, or core blocks, which could be rapidly expanded if a war breaks out, and also to maintain some knowledge of the state of the art—to include a familiarity with modern high-technology equipment.

However, although the precise contingencies for which preparations should be made might be unclear, there has not been much disagreement over the view that Australia must have the capability to destroy an invading force at sea (or in the air) long before it reaches the country's shores. A good deal of additional procurement, therefore, is going to the navy; in the air force also, the main emphasis is on maritime strike capacity.

IX. Multinational arms production

Technological complexity, high costs and lack of weapon standardization are the main driving forces underlying the growing trend to co-production in the defence sector. This process is mainly a Western affair; WTO weapon inventories are, with a few exceptions, standardized on Soviet equipment. This section concentrates on Western industrialized countries, since arms co-production normally involves these countries. There is, however, also a new tendency towards increasing co-production between industrialized and Third World countries as well as among Third World countries themselves.

The overriding rationale for co-production agreements is related to military efficiency and financial advantage. It is therefore not surprising that the majority of agreements concern combat aircraft and missiles. These weapons are in general more complex and expensive than warships and armoured vehicles. Consequently, larger savings can be made through co-production. The most successful venture so far is the Euromissile consortium set up by Aerospatiale of France and Messerschmitt-Bölkow-Blohm of FR Germany to develop anti-tank missiles and surface-to-air missiles such as HOT, MILAN and Roland. The consortium will also, with the participation of British Aerospace (BAe), develop the new NATO short-range air-to-air missile designated ASRAAM.

Examples of jointly produced aircraft include the Tornado and Jaguar fighters and the Franco-German Alpha Jet trainer. Various plans for a new European aircraft—the ECA and TKF projects—were abandoned in 1981 due to lack of funding and divergent opinions on what type of aircraft is required. Also in 1981, Westland and Augusta signed a second Memorandum of Understanding for joint development of a new ASW helicopter designated EH-101. Turning to warships, it has been estimated that the duplication costs in research and development within the NATO shipbuilding programme equal 20 to 30 new frigates a year [28]. An authoritative observer commented on the state of NATO warship standardization, as follows: "There is no doubt that NATO standardization is a mess . . . Minor alleviations have taken place, but national industrial demands have taken precedence over the requirements of the alliance and, in a rapidly advancing technological environment, electronic standardization is chaotic" [29].

The attitude of the United States towards arms co-production has been changing over the last couple of years. The USA has traditionally been very protectionistic in this matter and has also considered European weapon technology to be inferior to its own. Some years ago, however, the RSI-concept was introduced by William Perry, at that time Under Secretary for Defense. RSI stands for rationalization, standardization and interoperability, the broad objectives for achieving greater effectiveness of Western defence through co-operative weapon programmes within NATO. The Reagan Administration is stressing company-to-company agreements aimed at developmental cost sharing, dual production and reciprocal procurement decisions. Examples of this approach during 1981 are: in August, an agreement was signed between McDonnell-Douglas and BAe to produce jointly some 400 AV-8B Harrier V/STOL aircraft for the US Marine Corps and the Royal Air Force; and BAe will also be involved in the production of a new advanced jet trainer for the US Navy and Air Force.

Nevertheless, defence collaboration between the United States and European NATO countries is still in a formative stage. The reason for the slow development of arms co-production seems to be that there is a disadvantage for every advantage. Industrial interests, national interests and alliance interests are seldom coherent and mutual; they tend instead to pull in different directions. From an industrial point of view, the high costs involved in developing and producing weapons demand huge financial resources and large production units. In this respect, the arms industries are subject to the same forces as industries producing civil products, where the trend towards multinational production is well known. Reasons such as these explain the decision by Aermacchi and Aeritalia of Italy and the Brazilian manufacturer EMBRAER jointly to develop and produce the AM-X strike fighter.

On the other hand, many arms industries are unwilling to part with the technology that they have developed. The advantages of avoiding duplication costs through sharing of research and development outlays are often countered with various performance problems when the weapon system is tested. It is also argued that the division of labour in research and development undermines a continued indigenous design capability. Economies of scale cannot be fully exploited in sales to third countries because of different arms export regulations in the countries producing the weapon. Logistic support and repair and maintenance capacity are obviously made easier by standardization, but the complex and often bureaucratic production organization may escalate costs to unacceptable levels. The Tornado programme, for example, with its 500 companies and 70 000 workers in the three participating countries, is so complex that a minor slow-down of work in one country immediately slows down that in the other two countries as well. Finally, arms co-production may require common operational requirements and common tactical concepts. These differ among the NATO allies [30].

In sum, the objectives of RSI meet with problems connected with national security, employment, technology and trade. There is as yet no firm political and multilateral foundation for joint production and procurement of weapons.

References

1. *Soviet Military Power* (US Department of Defense, Washington, D.C., 1981).
2. Halbert, G. A., 'World tank production', *Armor*, March/April 1981.
3. Fallows, J., 'Arming for self-destruction', *New York Review of Books*, 17 December 1981.
4. *Report No. 97–58 to accompany Department of Defense Authorization for Appropriations for Fiscal Year 1982*, Committee on Armed Services, US Senate (US Government Printing Office, Washington, D.C., 1981).
   (a) —, pp. 25–26.
   (b) —, p. 36.
5. Stockman, D., quoted in *International Herald Tribune*, 15–16 November 1981.
6. Yankelovich, D. and Kagan, L., 'Assertive America', *Foreign Affairs*, Vol. 59, No. 3, 1981, pp. 696–713.
7. Weinberger, C. W., Statement to the Committee on Armed Services, US Senate, 4 March 1981 (US Government Printing Office, Washington, D.C., 1981), p. 3.

8. Weinberger, C. W., luncheon address to UPI News Agency in Chicago, 5 May 1981, *The Times*, 6 May 1981.
9. Thurow, L. C., *New York Times*, 31 May 1981.
10. Furham, R., Testimony to the Defense Industrial Base Panel of the House Committee on Armed Services, quoted in *Council on Economic Priorities Newsletter*, May 1981.
11. Thurow, L. C., 'How to wreck the economy', *New York Review of Books*, Vol. 28, No. 8, May 1981.
12. *Sixth Report from the Expenditure Committee, Session 1978–79*, The Future of the United Kingdom's Nuclear Weapons Policy, House of Commons Defence and External Affairs Sub-Committee (HMSO, London, 1979), p. 229.
13. *The Economist*, 10 October 1981.
14. Hennessy, P. and Greenwood, D., 'Uncovering the real defence cuts', *The Times*, 7 July 1981.
15. *Defense Business*, 11 October 1981.
16. Würtz, P., quoted in *Milavnews*, May 1980.
17. Würtz, P., quoted in *Atlantic News*, 17 December 1980.
18. Interview in *Le Monde*, 15 July 1981.
19. *Defense of Japan* (Japanese Defense Agency, Tokyo, 1981).
    (a) —, p. 147.
    (b) —, p. 301.
20 *Defense Daily*, 24 December 1981.
21. Okabe, N., 'Defense debates sharpen among business leaders', *The Japan Economic Journal*, 8 September 1981.
22. *Congressional Quarterly Weekly Report*, 10 October 1981.
23. *Financial Times*, 22 April 1981.
24. Middleton, D., *New York Times*, 14 March 1980, p. 11.
25. Zawawi, Q. A. M., Omani Minister of State for Foreign Affairs, quoted in *Financial Times*, 13 November 1981.
26. *Newsweek*, 19 September 1980.
27. *Africa*, January 1981.
28. *Financial Times*, 16 July 1981.
29. Foreword to *Jane's Fighting Ships, 1981/82* (Macdonald & Co., London, 1981).
30. *Financial Times*, 5 September 1981.

Appendix 2A

An illustration of weapon development: the Maverick and the Condor

Square-bracketed numbers, thus [1], *refer to the list of references on page* 61.

I. Introduction

The purpose of this short study is to take one particular area of weapon development as an illustration of the process. Two air-to-surface missiles—the Maverick and the Condor—have been selected from among the very large number of missiles which have been developed. The study concentrates on the development of one particular component of these missiles—the guidance system. The story of the development of these missiles illustrates a number of aspects of the process: the constant search for product improvement, inter-service rivalry, and the pressure from contractors for the continuation of projects once they are begun.

Certainly, so far as air-to-surface missiles are concerned, a good part of research expenditure has been concentrated on the development of new techniques of guidance. The first method of steering a missile was by radio command. This method was developed in Germany and the United States during World War II, when several missiles came into operation. Towards the end of the 1950s a new generation of radio command missiles, such as the US Bullpup AGM 12A and the French AS.20, came into operation. Bullpup AGM 12A was followed by another five versions (the last one became operational in 1970) and the AS.20 by several other French radio command missiles. Originally the missile had to be steered by keeping it on a sightline to the target using a radio command 'joystick'. This method was successively refined; in some versions the operator was freed from the need to align the target with his sight. Radio command is also a common technique of guidance now in use for Soviet air-to-surface missiles against land and sea targets; these missiles were taken into service from the end of the 1950s [1].

Radar is another method which was developed for steering missiles and is now one of the most widely used means. One mode, called active radar, is that in which the missile sends out a radar beam whose reflection reveals the location of the target; this is a common guidance technique against ship targets, which was first used for the US Bat glide bomb. The first missile equipped with this type of radar was the Swedish RB04, which was put into service in 1958. Other missiles using this mode were deployed during the 1970s. A missile can also be equipped with a radar seeker, to home on enemy radars. This is the passive radar guidance mode. The first anti-radar missile to enter service, the US Shrike, was introduced in 1964 and some versions are still operational. The Shrike was followed by the Standard ARM and the HARM; the HARM is still in development. The Soviet Union has developed the AS-X-9 anti-radar missile [1].

Some of the newest guidance systems now being developed use millimetre-wavelength radar, in particular for missiles directed against armoured targets. Ways have been found of overcoming the disturbance provided by smoke and dust. The seeker works initially as an active radar to search for appropriate targets and lock on to one. As the missile nears its target, 'glint' from multiple reflecting surfaces, could cause guidance problems; the seeker then switches to its passive mode, to home on millimetre-wave energy from the sky that is reflected from the target [2].

Three other methods of guidance are by a television camera fitted to the weapon, by a laser beam, and by infra-red or heat-seeking guidance. These are all examined in more detail in this study in connection with the Maverick and the Condor missiles, for in the process of the development of these missiles all three techniques have been employed.

The method of fitting a TV camera to the weapon to assist in guidance was first tried during World War II by the USA and Germany. Glide bombs equipped with a camera transmitted a TV picture to the aircraft operator, who in turn gave commands back to the bomb by radio. These bombs were thus not 'launch and leave' weapons. Only one bomb became operational (the US GB-4), and it did not work well. At the end of the 1960s the Walleye and the Electro-Optical Glide Bomb (EOGB) became operational. These bombs homed on their own to the target by a TV camera fitted to the nose of the weapon. TV guidance was the method of guidance used in the Condor and the first method used in the Maverick. The latter indeed appears to be the only air-to-surface missile using this form of guidance which has become operational.

Missiles can also be guided by slaving them to a laser beam which is illuminating the target. This method is at present in development for the Maverick as well as for the US Hellfire (an anti-armour missile whose principal launch platform is the helicopter), the Soviet AS-X-10 and the French AS.30L. Finally, there is the infra-red, or heat-seeking guidance technique. This was initially developed particularly with sea targets in mind, because at sea there are fewer problems in discriminating between targets. However, the Imaging Infra-Red (IIR) technique is also being developed for the Maverick for use against both ground and sea targets. It is also being developed for the US Harpoon anti-ship missile, and some foresee this type of guidance also for the French AS.30 missile.

II. The Condor [3]

The great disadvantage of radio command techniques of guidance is that the aircraft has to stay nearby until the missile strikes the target; thus it is vulnerable to enemy defence. The radio command mode has the further disadvantage that it is open to the enemy's electronic countermeasures (ECM). It is obviously preferable to have a type of guidance which is self-homing—that is, without the need to have any data links to the aircraft. The aircraft can then 'fire and forget'. It is free for evasive manoeuvres after launch.

The search for alternative, more satisfactory modes of missile guidance began in the 1960s. There was "an explosion in the technology of tactical weaponry, especially in air-to-surface missiles" [3]. It was not only the US Air Force which began to explore the possibilities of TV guidance: the US Navy did so as well. The Air Force began experimental work with the 'Hornet' programme in 1963 and the development of the Maverick in 1965; the Navy began the development of its Condor missile in the same year. Both were to use TV techniques.

The Navy's original requirement was in some ways similar to that of the Air Force. It was also for a stand-off guided missile to allow pilots to stay clear of enemy defences. However, the Navy wanted a missile with a longer range which would be capable of destroying heavily defended, high-value targets. Its primary purpose was to destroy land targets such as bridges, power stations and dams; the secondary mission was against ships. The Condor's range was usually estimated to be between 60 and 90 km. The contract for the Navy missile was given to Rockwell International.

As early as 1968, the first of many efforts was made to terminate the Condor programme; this first termination proposal came from the Assistant Secretary of Defense for Systems Analysis. There were many other critics in the course of this missile's history—in the House Appropriations Committee, the General Accounting Office, the Office of Management and Budget, and several sectors of the Office of the Secretary of Defense. The burden of the criticism lay in such matters as cost increases, programme delays, and the availability of other, much cheaper weapons, such as the electro-optical glide bomb and the Walleye—both of which could deliver the equivalent of a 900-kg bomb, as against 286 kg for the Condor missile.

The Air Force was constantly offered the Condor missile, and constantly refused it. The Air Force claimed that when slant range was considered, Condor's capability was not adequate to place the aircraft outside the range of hostile surface-to-air missiles. There were also questions about Condor's mid-course guidance by data link—as to whether this was or was not unjammable.

In spite of these criticisms, and in spite of the number of tests and evaluations in which the Condor failed to meet its performance targets, the missile was kept in the Defense budget for a further eight years from 1968. It was, of course, strongly defended by the Navy. Together with the Navy, Rockwell International—in particular, its Missile Marketing Division—lobbied intensively to keep the missile in development and to move it on to the production stage. Rockwell International had good reason to be concerned. It had suffered a prolonged slide in funding from the Department of Defense from the early 1960s to the early 1970s, mainly because it had lost business in the aircraft field and had failed to make gains in missile production. Condor was therefore very important to Rockwell; it was not the prime contractor for any other missile system that was either in production or near that threshold. Rockwell's lobbying included, *inter alia*, entertaining a number of Defense Department and Navy officials concerned with Condor at a hunting lodge in Maryland.

However, by 1976 the estimated unit cost of Condor, which had been put at $70 000 in the early stages of its development, had risen to $415 000 (in 1976 dollars); by 1975 the total estimated programme cost had reached $412 million. Cancellation eventually came in 1976.

III. The Maverick [1, 4–6]

The development of the TV Maverick missile was less troublesome. The initial contract to Hughes Aircraft was for a total package in which development, testing and all the elements that make up a complete operational system are bought together. Hughes Aircraft claims that Maverick is the only weapon system developed under the total package arrangement which did not overrun in cost. Development started in 1965. The first fully guided flight was made in 1969 and the first version of the missile, designated AGM 65A, was operationally deployed in 1973. The success of this missile has led to a series of further versions, designated AGM 65, versions B to F; their characteristics are summarized in table 2A.1 and they are described below.

In these further developments, the main changes have been in the guidance system. Other parts of the missile have remained much the same. The airframe in all versions is 2.46 m long, with a diameter of 0.30 m. The engine for the AGM 65A–D is the Thiokol TX-481 with dual thrust using a solid propellant and giving a speed greater than Mach 1. The E and F versions use a new propulsion unit which is only a modification of this engine with a new smokeless propellant. The weight depends on the

Table 2A.1. **The Maverick missile: characteristics of successive versions**

Designation	Military service	Guidance	Weight (kg)	Warhead	Weight of warhead (kg)	Status	Year of operational deploy-ment
AGM 65A	AF	TV	210	Conical-shaped charge	59	Operational	1973
AGM 65B	AF	TV scene magnifi-cation	210	Conical-shaped charge	59	Operational	1976
AGM 65C	AF, MC	Laser	210	Conical-shaped charge	59	Cancelled	–
AGM 65D	AF	Imaging infra-red	210	Conical-shaped charge	59	In develop-ment	1983 (est.)
AGM 65E	MC	Laser	286	Blast/fragmen-tation	135	In develop-ment	1983 (est.)
AGM 65F	N	Imaging infra-red	286	Blast/fragmen-tation	135	In develop-ment	1984 (est.)

AF = Air Force, MC = Marine Corps, N = Navy
Conical-shaped charge warhead = a warhead whose forward face has the form of a deep re-entrant cone; upon exploding, this directs a jet of gas and vaporized metal forward at such a speed that it penetrates thick armour.
Blast/fragmentation warhead = a warhead which relies on both blast and the fragments of a thick-walled casing or rod.
Source: References [1, 4–6].

warhead, which is either a conical-shaped charge or a blast/fragmentation warhead.

The Maverick is launched from rail launchers underneath the aircraft's wings in clusters of three or singly. It is operational on many types of aircraft, including the F-4, A-10, F-5, A-7 and the AJ37 Viggen. It is also intended for use on several other types.

The 'flight mechanical' range is given by the propulsion system and is almost the same for all versions. This range depends very much on the launch speed and altitude. For a low-level launch the range is estimated at 10–15 km. Launches from high altitude (10 000 m) will give ranges up to 40–50 km.

The *actual* range depends on the possibility of acquiring the target and locking on to it; this means that target size and scenario have a significant influence, as does the aircraft target acquisition system. For example, a tank in open terrain can be acquired at about 5 km with the A version and at about double that range with the B version. The C to F versions are more likely to use different acquisition systems in the aircraft, which means still longer detection ranges [7].

Maverick has many roles. Apart from the air-to-ground role, it can also be used against ships. In its ground attack role it is intended for use against small hard targets such as tanks, armoured vehicles, field fortifications, gun positions, concrete communications centres and aircraft shelters.

IV. TV-guided Maverick (AGM 65A and B)

The TV-guided Maverick works in the following way. While still attached to the underwing rail launchers, the Maverick through its nose-camera presents a view of what it sees to the pilot on a TV screen in the cockpit. The pilot locates the target on the display, moves the missile camera so that the fixed set of cross-hairs lies over the target, gives the lock-on command and launches the missile. The missile continues to the target on its own, guided by the target image keeping the lock-on gate on the target until impact. The pilot is free to veer away or attack other targets.

The Air Force thus acquired a weapon which increased the survivability of the aircraft (compared to missiles guided by radio command) by reducing the time spent in an exposed position. It also eliminated the threat of ECM interference. However, this missile could only be used during daytime and in good weather conditions. This also meant that the missile was vulnerable to such countermeasures as smoke.

Compared to the TV-guided glide bombs—the EOGB and the Walleye—the Maverick missile has the advantage that for any given range it can be launched from a lower altitude; the aircraft will thus be less exposed to an enemy radar.

In all, 13 000 rounds of the AGM 65A were produced for the USAF until production stopped in 1976. Another 6 000 rounds were produced for export. The missile has been sold to Egypt, Greece, Iran, Israel, Saudi Arabia, South Korea, Sweden and Turkey. As a result of the Swedish government's decision to buy the Maverick, the Swedish RB05B was discontinued in 1977. Israel, after its purchase of the Maverick, probably discontinued development of its own TV-guided missile, the Luz.

The next version of the Maverick—the AGM 65B—was developed during the 1970s and operationally deployed in 1976. It works in the same fashion as the AGM 65A: the pilot acquires the target visually through a TV picture presented on a display in the aircraft. The missile is locked on to the target before release and homes on to it without command from the aircraft.

The essential difference between the A and the B versions is the introduction in the B version of the scene magnification seeker. The field of vision was diminished from 5° to 2.5°, and at the same time the TV picture of the target was considerably magnified. This means that the range was increased (some estimates say it was doubled), or that much smaller targets could be attacked. The increased range in its turn increases the chance of survival of the aircraft.

Around 7 000 AGM 65Bs were initially produced. None of the initial production was exported. However, foreign military sales of the A version led to an increased demand for the B version to replenish home inventories [8]. In 1981 the production line was reopened to fulfil some overseas orders for the AGM 65B. Switzerland and Singapore have requested permission to buy this version. Some small changes have been made: the propellant is now near-smokeless, which of course increases the difficulty for ground-to-air defences.

Hughes Aircraft claims considerable operational accuracy for the Maverick AGM 65A and B versions. However, it has been argued that tests have not been carried out under realistic conditions. Opinions also differ on whether the missile failed or was successful when used in Viet Nam. The development costs of the AGM 65A and B were $144.7 million [9]. The average unit historical costs of the two versions (excluding those produced from the re-opened production line in 1981) came to $16 000 per missile [10]. The current cost, at 1981 prices, is of the order of $50 000 per missile [11]. These are the unit costs of one completely equipped missile. No spares are included. The price at which the missile is sold is higher.

V. The laser Maverick (AGM 65C and E)

The disadvantage of the TV-guided system is that it can only operate in daylight and in good weather. The next stage in the development of guidance systems for the Maverick missile was to look for methods which avoided these disadvantages. The use of a laser beam was one such method.

The first flight of a missile with laser guidance was made in 1965. In 1968, Paveway bombs, using a laser, became operational and were used in Viet Nam. Then between 1969 and 1971 the Marine Corps sponsored the development of a laser-guided missile, the Bulldog.

A laser-seeking system means that the missile is equipped with a laser seeker which homes on to an illuminated spot. This spot is imposed on the target by a laser designating device. This designator can be installed in the missile-carrying aircraft itself or in another aircraft nearby, or it can be held by a person on the ground close to the target. The Marine Corps proposes to use the last of these three methods, with a soldier on the ground designating a target which he wants destroyed; the laser-guided missile fired from the aircraft would then home onto that target, with the aircraft pilot possibly not knowing what the target was.

The laser Maverick is superior to the TV Maverick in several respects. Apart from its day/night capability, it is also to be preferred when attacking low-contrast and unbounded targets (i.e., targets lacking well-defined visual contrast features). Laser countermeasures such as smoke, while a problem, are operationally impractical to use in a battlefield environment where the missile would be employed [12a].

The development of a laser-guided missile in the United States was complicated by the competition between the Air Force's Maverick missile and the Marine Corps's Bulldog. In 1971 the Air Force was given the task of developing a laser seeker which could be used by all three services—the Army, the Marine Corps and the Air Force. It was to be used on the Army's helicopter-launched anti-armour missile, the Hellfire, on the GBU 15 glide bomb, and on a missile which was to be common between the Air Force and the Marine Corps [13].

The decision on the common missile was preceded by arguments from the Air Force and the Marine Corps in favour of their respective missiles. The Marine Corps claimed that the Bulldog would have a unit programme cost of around $21 000 while the seeker of the Maverick alone would cost $28 750 per copy. The Air Force stated that their Maverick seeker would cost only $5 000 and the whole missile $13 500 [14]. The decision went in favour of the Maverick, and in 1974 the Bulldog programme was cancelled. However, during the long period in which the laser Maverick was being developed, the Defense Department did from time to time reconsider the Bulldog as an interim measure for the Marine Corps.

In the course of development of the laser-guided missile, problems arose particularly with the aircraft designator, and in any case the Air Force became more interested in the imaging infra-red guidance system. In 1979 the Air Force withdrew its demand for a laser-guided missile, leaving the Marine Corps as the sole customer. Since the Marine Corps intended in any case to use a ground-based laser designator, it was not particularly concerned with the fact that there were difficulties with aircraft designation. The description of the laser-guided version of the Maverick was then changed from AGM 65C to AGM 65E (see table 2A.1). The Marine Corps wished to have the heavier blast/fragmentation warhead, since it intended to use it against targets requiring these larger warheads.

The laser Maverick has become an expensive missile for the Marine Corps. Whereas the development costs for the Bulldog were only $16 million [12b], they were $65.3 million for the laser Maverick [9]; these costs, now that the Marine Corps is the sole customer, are to be spread over a fairly small number of missiles. The Corps has requested 4 600 missiles. The missile should become operational in 1983.

The common tri-service laser seeker, which the Air Force had been given the task of developing in 1971, has not as yet materialized. The GBU-15 glide bomb laser-seeker programme is terminated, and the Hellfire missile development continued with a seeker which the Army claimed was less expensive. However, that claim is now being questioned, and the Senate Armed Services Committee has requested the Secretary of Defense to "assess the possibility of using the Maverick Tri-Service seeker on the Laser Hellfire" [15].

VI. The IIR guidance Maverick (AGM 65D and F)

From the Air Force's point of view, the laser Maverick had a number of disadvantages. It was not a 'launch and leave' weapon, the aircraft designator tests had been unsatisfactory, and, although the laser system could cope with 'normal' rain, it could not cope with fog or smoke. The Air Force therefore concentrated its efforts on development of the imaging infra-red guidance system.

The infra-red seeker forms a TV-like picture by sensing the difference in infra-red heat radiated by objects in view. It has many clear advantages over the TV and laser versions. It is a 'launch and leave' weapon, it has been claimed to have up to three times the lock-on, tracking and launch-range of the TV Maverick, and it is an adverse-weather, night or day system. It can penetrate battlefield smoke and dust. It can be used against hidden or camouflaged targets and can distinguish decoys by their low temperature. However, it cannot be used in heavy fog or heavy rain, since the humidity would absorb the infra-red heat.

The infra-red seeker is also more sensitive than the laser seeker. However, when the centre of a target is not the best place to hit, an IIR missile can be used together with a laser spot tracker to guide the missile to the most vulnerable spot, which is then illuminated by a laser designator.

The development of the IIR seeker began as early as 1970, but it was not until 1974 that it reached the advanced development stage. In 1976 Congress denied further funding, because it was doubtful whether such a development would be cost effective. The programme was reinstated in 1977, and in that year Hughes Aircraft was awarded a contract to define a common IIR seeker assembly for the GBU-15 glide bomb and the Walleye II guided bomb as well as the Maverick [16].

The AGM 65D (see table 2A.1) will be procured by the Air Force and is expected to enter service in 1983 [17]. The normal review which precedes a full-scale production decision has been set aside for this missile, as it has for other high-priority programmes. The Air Force plans to procure 60 000 rounds of this missile. The Navy is proposing to procure a slightly different version, the AGM 65F. In this version the tracker's software is specially programmed for ship attack, and this model is fitted with the blast/fragmentation warhead. The fuse delay can be selected to allow the air crew to choose the best setting for detonation inside the target ship. The Navy plans to buy 7 000 of the AGM 65F, to be delivered by mid-1984 [18].

The development costs of the IIR Maverick have amounted to $185 million up to the end of FY 1981 [9]. The unit cost of AGM 65D and F is estimated at $75 000, in 1983 dollars.

All three versions of the Maverick missile are therefore going into production. Hughes Aircraft is preparing to produce around 200 a month of the TV version, between 100 and 200 a month of the laser version, and 500 a month of the IIR version.

VII. Conclusions

This short study of the Maverick missile illustrates some aspects of the process of weapon development. It illustrates, for example, the long lead-times in this process. The development of the IIR seeker began in 1970; however, an air-to-surface missile equipped with such a seeker will not enter the inventory until 1983.

The process of development of the guidance system for air-to-surface missiles had two main objectives. One was for a 'launch and leave' capability, from as great a range as possible, so that a missile could be fired from a position beyond the defensive capability of the target and so that the aircraft could then immediately leave the scene after firing the missile. The second main objective was for an all-weather day and night system. The TV Maverick was a 'launch and leave' missile. Whether, however, it could be fired from a position beyond the defensive capability of the target would depend partly on how that target was defended. Further, it was only a daylight/fair-weather missile.

The laser-guided version had the advantage over the TV version that it could be operated at night and in light or 'normal' rain. However, it had to have a laser designator. As at present envisaged, it will be a system by which a soldier on the ground can call in an air strike on a target which he has designated.

The IIR version has the advantage over TV Maverick both in range and in the fact that it can be operated both by day and night and also in poor weather. Its use can be combined with terminal laser guidance, to direct it to a particular spot on the target. It is also a 'launch and leave' missile. So the process of development, perhaps at a total cost of the order of $500 million, has brought the air-to-surface missile much nearer to the objectives originally set.

The account of the development of Maverick also illustrates another theme common to weapon development: inter-service rivalry and the opposite trend to a search for commonality in weapons between the three services. The Navy persisted with the development of the Condor missile long after the evidence suggested that it could not be cost effective. There was an attempt to get a tri-service laser seeker which could be used on weapons deployed by all three services. The attempt appears so far to have been unsuccessful. On the other hand, there does seem to have been some success in developing an IIR seeker which is common to more than one of these services, and there is also now the common Air Force–Navy programme in the development of the IIR Maverick missile.

Postscript

On 23 February 1982, the Pentagon announced that it will hold up full-scale production of the IIR (AGM 65D) Maverick missile until technical problems are ironed out, leading to a delay in the programme. It appears that some of the missile's components, including the seeker, have failed in testing. Recent tests have shown problems in identifying the target: a sun-heated rock could be taken to be an enemy tank.

There have also been suggestions that earlier tests of the IIR Maverick have been unrealistic. The target tanks moved in a way that the pilots could predict. Even under these conditions, the test results were not impressive. However, a spokesman for the United States Air Force stated that the IIR Maverick is still a high-priority item in their weapon procurement list [19].

References

1. Gunston, B., *Rockets & Missiles* (Salamander Books, London, 1979).
2. *Aviation Week & Space Technology*, Vol. 111, No. 3, 16 July 1979, p. 59.
3. *Conflict of Interest and the Condor Missile Program*, Report by the Subcommittee on Investigations of the Joint Committee on Defense Production, US Congress (US Government Printing Office, Washington, D.C., September 1976).
4. *Jane's Weapon Systems, 1980–81* (Macdonald & Co., London, 1980).
5. *Flight International*, Vol. 119, No. 3760, 30 May 1981, p. 1620.

6. Gervasi, T., *Arsenal of Democracy, American Weapons Available for Export* Grove Press, (New York, 1977), p. 184.
7. Personal communication with Head of Division K.-O. Andersson, Swedish Defence Materiel Administration.
8. *Hearings on Military Posture and H.R. 5068* [*H.R. 5970*] *Department of Defense Authorization for Appropriations for Fiscal Year 1978*, before the Committee on Armed Services, House of Representatives, 95th Congress, 1st Session (US Government Printing Office, Washington, D.C., 1977), Part 2, p. 550.
9. *World Missile Forecast* (Forecast Associates, Ridgefield, Conn.), Maverick Entry of April 1981, p. 4.
10. *Jane's Weapon Systems, 1979–80* (Macdonald & Co., London, 1979), p. 146.
11. Geddes, P., 'Maverick missile enters new phase', *International Defense Review*, Vol. 14, No. 11, 1981, p. 1466.
12. *Department of Defense Appropriations for 1980*, Hearings before a Subcommittee of the Committee on Appropriations, House of Representatives, 96th Congress, 1st Session (US Government Printing Office, Washington, D.C., 1979).
    (a) —, Part 7, p. 390.
    (b) —, Part 7, p. 391.
13. *Aviation Week & Space Technology*, Vol. 99, No. 19, 5 November 1973, p. 56.
14. *Aviation Week & Space Technology*, Vol. 99, No. 24, 10 December 1973, p. 44.
15. *Department of Defense Authorization for Appropriations for Fiscal Year 1982*, Report, Committee on Armed Services, United States Senate (US Government Printing Office, Washington, D.C., 1981).
16. Howe, R. W., 'Air-to-surface weapons for the tactical fighter force', *Pacific Defence Reporter*, Vol. 6, No. 5, November 1979, pp. 75–76.
17. *Interavia Air Letter*, No. 9680, 3 February 1981, p. 3.
18. *Flight International*, Vol. 18, No. 3736, 13 December 1980, p. 2164.
19. *Washington Post*, 23 February 1982, p. 1.

Appendix 2B

World military expenditure, 1972–81

Table 2B.1. World military expenditure summary, in constant price figures

Figures are in US $ mn, at 1979 prices and 1979 exchange-rates. Totals may not add up due to rounding.

	1972	1973	1974	1975	1976	1977	1978	1979	1980	1981
USA	134 794	127 972	126 514	122 688	116 045	120 805	121 595	122 279	126 865	134 390
Other NATO	81 684	83 174	85 753	87 837	89 672	91 204	94 393	96 282	98 546	99 567
Total NATO	216 478	211 146	212 267	210 525	205 717	212 009	215 988	218 561	225 411	233 957
USSR	[103 900]	[105 600]	[107 300]	[109 000]	[110 400]	[112 100]	[113 700]	[115 200]	[116 900]	[118 800]
Other WTO	8 993	9 420	9 869	10 612	11 061	11 461	11 798	11 985	12 100	[12 795]
Total WTO	[112 893]	[115 020]	[117 169]	[119 612]	[121 461]	[123 561]	[125 498]	[127 185]	[129 000]	[131 595]
Other Europe	10 872	11 101	11 883	12 404	13 012	12 962	13 079	13 612	13 754	(13 627)
Middle East	12 320	17 249	24 909	30 784	34 037	33 043	33 432	(34 918)	(36 396)	[43 950]
South Asia	4 601	4 154	3 969	4 356	4 931	4 774	4 969	5 037	[5 480]	[5 587]
Far East (excl. China)	14 795	15 800	16 010	18 167	20 133	21 547	23 811	24 569	25 088	26 654
China	[29 000]	[30 700]	[30 700]	[32 400]	[33 200]	[32 300]	[37 000]	[44 400]	[42 700]	[37 200]
Oceania	2 995	3 074	3 326	3 429	3 417	3 430	3 428	3 446	3 617	3 906
Africa (excl. Egypt)	6 519	6 795	8 525	(10 219)	(11 250)	(11 358)	(11 468)	[11 690]	[12 450]	[13 600]
Central America	1 094	1 152	1 226	1 366	1 647	1 950	2 085	2 158	2 134	2 299
South America	4 737	4 854	5 645	5 159	7 240	7 193	7 249	7 351	[6 512]	[6 352]
World total	**416 304**	**421 045**	**435 629**	**448 421**	**456 045**	**464 127**	**478 007**	**492 927**	**502 542**	**518 727**
Developed market economies	238 596	235 134	236 807	236 498	232 980	239 751	244 031	247 718	253 541	262 137
Centrally planned economies	[143 805]	[147 633]	[150 204]	[154 759]	[157 869]	[159 407]	[166 415]	[175 844]	[176 125]	[173 652]
OPEC countries	11 127	13 833	22 381	29 190	32 934	31 824	33 534	(34 120)	(37 400)	[46 220]
Non-oil developing countries:										
with (1978) GNP *per capita* < US $300	6 333	5 642	5 632	(5 957)	(6 390)	(6 050)	(6 576)	[6 740]	[7 284]	[7 687]
with (1978) GNP *per capita* US $300–$699	4 695	6 539	7 182	7 850	8 002	8 703	7 306	(7 008)	(6 850)	[7 090]
with (1978) GNP *per capita* > US $699	10 825	11 282	12 429	13 037	16 620	17 053	(18 669)	(19 970)	(19 782)	(20 284)
Total non-oil developing countries	21 853	23 463	25 243	26 844	31 012	31 806	32 551	33 718	33 916	35 061
Southern Africa	1 326	1 508	1 975	2 407	2 815	3 190	3 305	3 260	[3 170]	. .

3. The trade in major conventional weapons

Square-bracketed numbers, thus [1], *refer to the list of references on page* 78.

I. Introduction

There is at the present time little enthusiasm for any multilateral restraint of the international trade in arms. The CAT (Conventional Arms Transfers) talks between the United States and the Soviet Union have not resumed for the past three years, and the European arms suppliers have shown little, if any, inclination towards participating in multilateral restraint efforts. In the present climate of tense relations between the two great powers, this situation is unlikely to change in the near future. Furthermore, economic incentives, particularly in the West European arms manufacturing countries but also in the USA and the USSR, are becoming increasingly important. International tension (exemplified by recent events in Afghanistan, Poland, the Middle East and Central America), national economic considerations, and competitive fears of losing market shares all make the prospects for the control and eventual elimination of the global arms trade look bleak. Indeed, as one observer puts it:

> To oppose such a development may well place one in the role of an existentialist character, struggling against a fate he knows to be inevitable; but if the proliferation of conventional arms is an undesirable prognosis, it is perhaps the only basis for a critical moral stance. [1]

The flow of arms: general trends

The flow of arms during the period 1979–81 is shown in figures 3.1 and 3.2. (All the tables and figures in this chapter are based only on *actual deliveries* of major conventional weapons.) The Soviet Union has passed the United States as the world's largest major-weapon exporting country during the period (figure 3.1). This is partly due to a substantial increase in Soviet arms exports to India and to countries in the Middle East and North Africa, and partly to a decline in US exports mainly resulting from the policy of unilateral restraint initiated by President Carter in 1977.

The 1979–81 Third World share of total arms imports is approximately 62 per cent (figure 3.2), compared to a share of 69 per cent for the period 1977–80. The long-term trends in the arms trade with the Third World are shown in tables 3.1 and 3.2. The total value, measured in constant prices, for every five-year period has approximately doubled compared

Figure 3.1. Shares of world exports of major weapons, 1979–81, by country

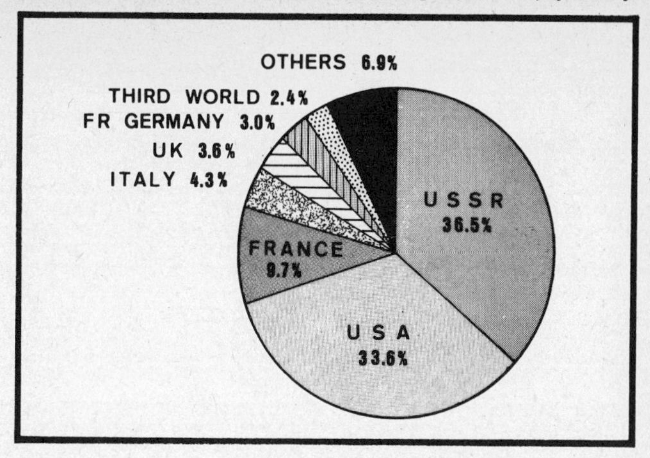

Figure 3.2. Shares of world imports of major weapons, 1979–81, by region

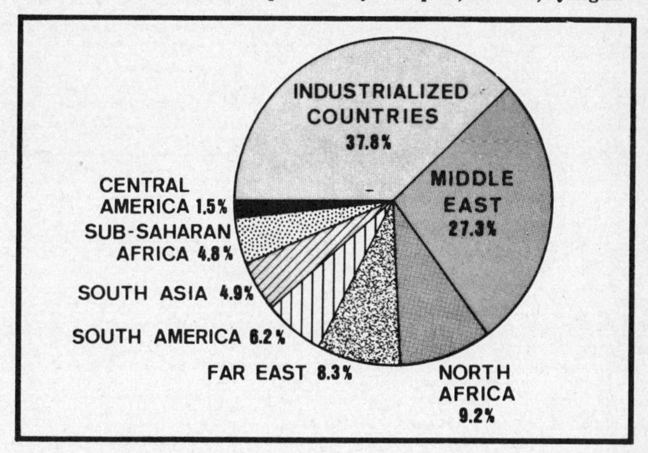

to the previous period. Among recipient areas, Africa has become more important. Among supplying countries, France and Italy have increased their share of total exports, and the UK share has fallen. (The growth of Third World arms imports on an annual basis is shown in figure 3.3.)

The increase in the world arms trade is both quantitative and qualitative. In the early 1960s, the vast majority of transferred weapons were relatively unsophisticated and second-hand. Today it is different. The current SIPRI arms trade registers—covering major weapons on order or being delivered in 1981—identify approximately 1 100 separate arms transfer agreements. Ninety-four per cent of these contracts are for new weapon systems, 2 per cent are for second-hand weapons, and 4 per cent are for refurbished weapons.

Table 3.1. Shares of imports of major weapons by the Third World: by region, 1962–81

Percentages are based on SIPRI trend indicator values, as expressed in US $ million, at constant (1975) prices.

Region[a]	1962–66	1967–71	1972–76	1977–81
Middle East	*28*	*46*	*51*	*44*
Africa	*15*	*9*	*16*	*24*
Far East	*31*	*27*	*15*	*13*
Latin America	*12*	*7*	*11*	*11*
South Asia	*14*	*11*	*7*	*8*
Total	*100*	*100*	*100*	*100*
Total value	**7 870**	**14 583**	**25 775**	**47 829**

[a] Regions are listed in rank order according to their shares for 1977–81.

Table 3.2. Shares of exports of major weapons to the Third World regions in table 3.1: by supplier, 1962–81

Percentages are based on SIPRI trend indicator values, as expressed in US $ million, at constant (1975) prices.

Country[a]	1962–66	1967–71	1972–76	1977–81
USA	*29*	*34*	*38*	*37*
USSR	*42*	*42*	*33*	*33*
France	*9*	*7*	*10*	*12*
Italy	*1*	*1*	*2*	*5*
UK	*12*	*10*	*9*	*4*
Others	*7*	*6*	*8*	*9*
Total	*100*	*100*	*100*	*100*
Total value	**7 870**	**14 583**	**25 755**	**47 829**

[a] Countries are listed in rank order according to their shares for 1977–81.

II. The suppliers

The United States

In May 1977, President Carter issued a directive outlining his policy on conventional arms transfers. The aim was to bring about a slowing-down in the international arms trade through a unilateral policy of US restraint, which in turn might lead the Soviet Union and other major suppliers to follow suit. Arms exports were only in exceptional cases to be used as

Figure 3.3. Imports of major weapons by the Third World, 1962–80

Based on five-year moving averages of SIPRI trend indicator values, as expressed in US $ million, at constant (1975) prices.

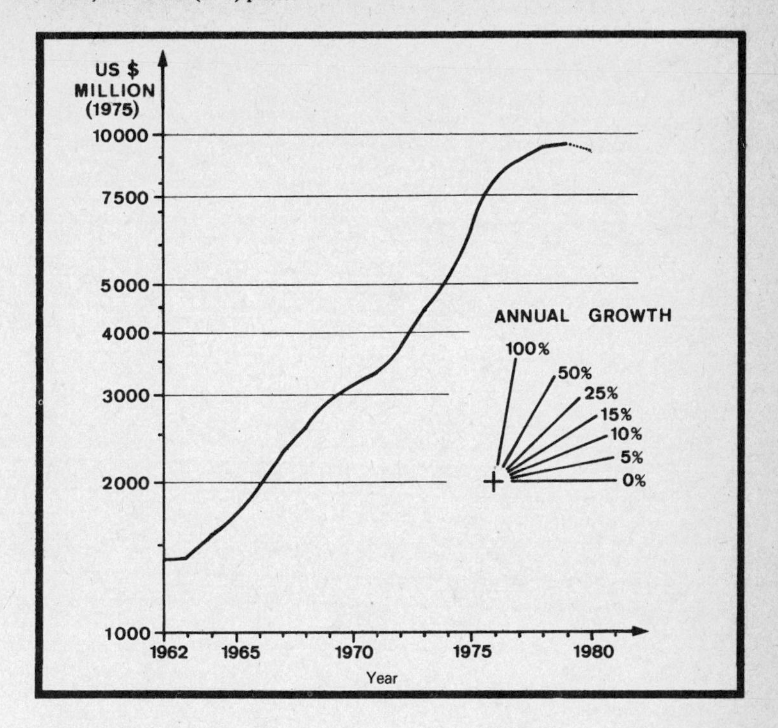

instruments of US foreign policy. The directive established a dollar ceiling for total US foreign military sales, and it indicated an intention not to introduce advanced weaponry that would significantly raise the combat capability in any given region. It restricted the resale of arms to third countries, and reaffirmed the link between human rights criteria and military assistance. Soon after this directive was issued, negotiations with the Soviet Union were begun on conventional arms transfers.

This was a praiseworthy attempt to curb the international trade in arms, but unfortunately it failed. Neither the Soviet Union nor West European arms suppliers were prepared to co-operate in a multilateral effort of restraint. Furthermore, from the very beginning, and particularly during 1979–80, the Carter Administration made several exceptions to its stated policy. A gap emerged between this policy and actual arms sales decisions.

A new policy

The restraints which remained were seemingly eliminated on 8 July 1981, when President Reagan signed a new Presidential Directive on arms transfers. The guidelines in this directive stem from the same philosophy that lies behind the rearmament programme described in the previous chapter: fundamental US interests are challenged by the Soviet Union and the stability in many regions considered vital to the USA is being threatened.

The following excerpt from the directive indicates the new attitude:

> The United States cannot defend the free world's interests alone. The United States must, in today's world, not only strengthen its own military capabilities, but be prepared to help its friends and allies to strengthen theirs through the transfer of conventional arms and other forms of security assistance. Such transfers complement American security commitments and serve important United States objectives. Prudently pursued, arms transfers can strengthen us. The United States therefore views the transfer of conventional arms and other defence articles as an essential element of its global defence posture and an indispensable component of its foreign policy . . . We will deal with the world as it is, rather than as we would like it to be. [2]

At this stage, the Reagan guidelines are more in the nature of a repeal of the Carter policy than the elaboration of a new one. The directive consists of a broad set of aims and principles rather than a specific set of rules. One thing, however, is made clear: the human rights issue is dead. The directive states than an important factor to be considered when making arms transfer decisions is "whether any detrimental effects of the transfer are more than counterbalanced by positive contributions to United States interests and objectives" [2].

The Reagan guidelines are basically a restatement of Republican advocacy in the early 1970s of the policy for arming Third World countries as a substitute for US military presence there. The so-called Nixon doctrine implied that the USA should help its friends and allies among the developing countries to help themselves. Measures to this effect, apart from cash sales, arms-for-oil agreements and so on, will be carried out through an extensive programme of military and economic assistance. The security assistance authorization for FY 1982 shows an increase of 30 per cent compared with FY 1981. Almost 70 per cent of the programme is intended for the Middle East; the bulk of it is allocated to Israel and Egypt. Other major recipients in the North Africa–Middle East area include Morocco, Somalia, Tunisia and North Yemen. A substantial part of the total security assistance consists of Foreign Military Sales (FMS) financing. To facilitate arms exports through the FMS programme, a number of changes have been proposed. A Defence Acquisition Fund will be established in order to procure equipment in anticipation of the requirements of allied and friendly nations. It is argued that this will minimize the damage to US force readiness that results from the diversion of US service stocks. It is also proposed to raise by a factor of two the reporting thresholds to Congress for transfers of weapons and other defence equipment [3].

It is perhaps not surprising that the Reagan Administration is more disposed than the Carter Administration to use arms sales as an instrument of foreign policy. The political arguments are coupled with economic ones. It is far cheaper, and less politically troublesome, to send arms abroad than to send US troops. A US soldier in Egypt, for example, would cost $150 000 a year, while an Egyptian soldier costs $2 100 a year [4]. Furthermore, arms transfers improve the foreign trade balance, provide some 800 000 jobs in the USA, and make US domestic arms procurement cheaper.

What is notable, however, is the Administration's tendency to consider arms transfers almost exclusively in an East–West context. A typical example of this is the wish to supply arms to several mutually hostile parties in the Middle East in an effort to contain Soviet penetration of the Gulf region. Such an approach may underestimate the driving forces underlying regional conflicts and may instead fuel local rivalries and arms races.

Applications of the policy

In January 1980, President Carter made a substantial exception to his declared policy when he allowed the production of a fighter aircraft, the FX, designed solely for export. The new Reagan guidelines also call for the production of equipment that more readily fits the needs of Third World countries in terms of cost and complexity. In practice, however, this principle has been undermined during 1981 by a number of decisions that imply dropping the distinction between first-class and second-class friends. The offer to sell 24 General Dynamics top-of-the-line F-16 fighters to Venezuela is one example of such a decision. Similarly, the long-standing request from South Korea to acquire new fighter aircraft has been approved by the Reagan Administration. At a cost of $900 million, South Korea will get an initial batch of 36 F-16s. This means the introduction into the region of weaponry of a significantly higher technological level than before, and it also means the risk of an intensified regional arms build-up. F-16s in the South Korean Air Force will probably induce the Soviet Union to provide North Korea with MiG-23 fighters. There is also the formidable $3.2 billion five-year military and economic aid package to Pakistan. The main item of this agreement is 40 F-16 fighters—possibly with some funding from Saudi Arabia—of which six will be delivered in 1981–82 by diverting European-produced F-16s to the US Air Force.

The package also includes attack helicopters, tanks, anti-tank missiles, artillery, armoured personnel carriers and advanced communications systems. Pakistan is considered part of the strategic line against the USSR that also includes the pro-Western Gulf states, Egypt, Israel and Turkey. In return, the United States is concerned to gain access to airfields and ports for the Rapid Deployment Force. The argument has also been put forward that the military aid package will help to prevent Pakistan from acquiring nuclear weapons.

Another notable example of a military assistance undertaking during 1981 is the US participation—together with Saudi Arabia, Egypt, Pakistan and China—in a clandestine operation to supply arms to the Afghan resistance. The weapons, mainly Soviet or replicated Soviet equipment from Egypt, include surface-to-air and anti-tank missiles as well as rifles and machine-guns [5]. Other recipients of significant US military assistance during 1981 include Egypt, El Salvador, Israel and Sudan. Furthermore, a request from the Administration for resumption of arms sales to Argentina and Chile has been approved by the Congress; this was previously prohibited because of human rights violations.

Two particular issues have come to the fore during 1981 in connection with US arms export policy. These are possible sales of arms to China and the $8.5 billion sale of five Boeing E-3A AWACS aircraft and other equipment to Saudi Arabia. During the Carter Administration it was decided that so-called dual-use technology and certain defence-related material would be made available to China. Some 400 export licences, including the export of electronics and military support equipment, were also granted to US companies. Very few, if any, of these sales were actually made. Then, in June 1981, US Secretary of State Haig said that the USA was willing to consider selling 'lethal' weapons to China on a case-by-case basis and in consultation with the Congress and allied countries. Restrictions on sales of military-related technology would also be drastically reduced. Among the weapons the USA is willing to sell are Hawk surface-to-air missiles, TOW anti-tank missiles and armoured personnel carriers. It is doubtful, however, if this offer will result in a major inflow of US weapons. China is wary of major foreign weapon purchases for a number of reasons. First, its financial resources are limited; defence has at present a relatively low priority and the emphasis on self-reliance in defence modernization means that China would rather buy fire control systems to improve the accuracy of existing tanks and missile systems than invest in advanced fighter aircraft it cannot manufacture. Second, the purchase of US weapons would create pressure on China to accept the sale of FX or even F-16 fighters to Taiwan as a *quid pro quo*.

From a US point of view this is a delicate question. The harsh Chinese attack on the Netherlands, including downgraded diplomatic relations, after the sale of two Dutch submarines to Taiwan obviously made an impression. Compromises are possible, but the conclusion is that the United States will continue to deliver defence-related technology rather than actual weapon systems. The so-called 'China card' is a powerful foreign policy instrument only so long as it is not played.

The AWACS sale is something altogether different. The agreement is perhaps the most important US arms transfer ever in terms of the money, the technology and the implications involved. Apart from the five AWACS surveillance and battle management aircraft, this air defence package comprises six KC-135 aerial refuelling tankers, 1 177 AIM-9L Sidewinder air-to-air missiles for 62 F-15 fighters already on order, long-range fuel tanks for the F-15s and 22 ground-based radar installations, 10 of which will operate with the AWACS planes. The opposition to this sale in the USA has centred around three arguments: first, the threat it poses to Israel; second, the risk that the sensitive technology could be revealed to the Soviet Union; and third, the risk that the identification of the Saudi regime with the US government might strengthen the position, within Saudi Arabia, of opponents of the existing Saudi government. The latter argument implies a development similar to that in Iran.

The Administration argues in favour of the deal on the grounds that it helps Saudi Arabia defend its oilfields against strike attacks from the Soviet Union or from pro-Soviet countries such as South Yemen or Ethiopia. It is also argued that the deal will help to restore US credibility as a reliable security partner; that the whole apparatus of training, logistics, support infrastructure, and so on increases US military presence in the region; and that the whole air defence system will be compatible with the equipment of US forces, thus facilitating the deployment of US soldiers and weapons to the region in time of need. Given the historical Saudi opposition to foreign military bases on their soil, the AWACS deal is the nearest thing to a prepositioned base structure that the USA is likely to obtain at the present stage. And, apart from the obvious fact that it is a cash sale that may lower the procurement cost for the US Air Force's own AWACS planes, this is the heart of the matter. Both the present and the previous US Administrations have, in co-operation with Saudi Arabia, been trying to create an integrated regional air defence system of US origin, led by Saudi Arabia. Defence collaboration within the recently formed Gulf Cooperation Council—including Saudi Arabia, Oman, Qatar, Bahrain, the United Arab Emirates and Kuwait—may be a further step in that direction. In its concentration on the East–West perspective, the US government has paid little attention to the possible alternative uses that Saudi Arabia might make of the package, to the internal consequences that might follow from the sale, and to the strong criticism of their closest ally in the region—Israel.

Other NATO

When an economy is in crisis, more weight is attached to economic than to political arguments. With high unemployment, foreign trade imbalances and budget deficits, this is now particularly evident in the West European arms manufacturing countries. Financial constraints have caused cuts, postponements and cancellations in most domestic defence procurement programmes. This has, in part, contributed to rising unit costs, thus inducing further cuts. For these reasons, the major arms producers are pushing military sales, particularly to Third World countries, more than ever before. Criticism is muted by strong national economic considerations: arms exports improve the balance of payments, lower unit prices through the advantages of scale, and ensure employment in the arms industries. As one French arms industry representative put it at the Satory defence exhibition: "if we don't export, in 20 years we'll be making propeller aircraft and wooden missiles" [6].

The French arms industry employs some 300 000 workers and is highly dependent on exports. More than 5 per cent of total French exports consist of weapons, and for the leading arms export company, the aircraft manufacturer Dassault-Breguet, exports constitute approximately 70 per cent of total turnover. French trade unions generally advocate arms exports for employment reasons and, together with the industry, they exert a major influence on public opinion and political decision makers. The attitude of French government officials towards arms exports is that it is up to the individual country to choose the weapons it will purchase; France should not interfere with the procurement policies of other countries by, for example, refusing to sell a certain weapon system. On the contrary, France should offer Third World countries a possibility to diversify their arms sources so that they need not become dependent on either the United States or the Soviet Union. The agreement with Nicaragua in December 1981—for two helicopters, two patrol boats and a training programme—illustrates this policy. Those in favour of restraint initially hoped that the Mitterand government would introduce a set of strict arms export regulations in accordance with campaign promises made. This, as it turned out, was not the case; the French government has evidently decided that the economic benefits which arms sales provide outweigh any moral arguments. However, 29 of 50 AMX-30 main battle tanks for Chile have recently been embargoed.

France exports weapons mainly to countries from which it receives something in return. In the year ending April 1981, the Middle East and North Africa took nearly 80 per cent of total French weapon sales [7]. In spite of the continuing war between Iran and Iraq, France delivered Mirage fighters to Iraq and missile-armed fast attack boats to Iran during

1981. Libya received weapons during much of 1981, with the exception of an export ban from February to July due to the Chad intervention. At the same time, France increased deliveries to the Central African Republic, Gabon, the Ivory Coast and Senegal in order to prevent Libyan aggression in these countries.

In the UK, arms exports provide jobs for 140 000 people and account for 2.5 per cent of all British exports [8]. The government regards arms exports as an important element in the eventual recovery of the British economy. Prime Minister Thatcher's April tour of the Middle East, following a visit to India, was the culmination of an intensive arms marketing effort conducted by strong British diplomatic and industrial teams. The main promotion item was the BAe Hawk, an advanced jet trainer/light strike aircraft. It is reported that the United Arab Emirates and Saudi Arabia soon afterwards signed contracts for 30 and 40 Hawks, respectively. Other British weapons destined for the Gulf states are Chieftain tanks and Rapier surface-to-air missiles. However, the British government has refused—despite reported requests from both sides—to supply arms or spare parts to Iraq or Iran while they are at war [9].

In July 1980, the government lifted the embargo on arms sales to Chile, which had been imposed in 1974 for human rights violations. During 1981, this resulted in a much criticized agreement to sell a missile destroyer of the County-class and a 27 000-ton fleet tanker to the Chilean Navy. Both ships were made redundant as a result of cuts in the Royal Navy surface fleet. The sale was defended by government officials on the grounds that the two ships could hardly be used in a counter-insurgency role [10].

In FR Germany there are problems of financing future defence outlays while at the same time the greater part of the West German arms industry is working at 50 per cent of capacity [11]. The question is whether FR Germany will openly follow France, the UK and Italy in their effort to export arms for economic reasons. The sale of two Kiel-Howaldtswerke Type 209 submarines to Chile has reportedly been stopped and the submarines have been offered to Denmark, but a large sale of armoured vehicles to Saudi Arabia is still pending. In order to go through, this deal —for Leopard tanks, Gepard anti-aircraft vehicles and Marder infantry combat vehicles—will require a substantial change in FR Germany's arms export policy. The policy of prohibiting sales to 'areas of tension', for example, will have to be revised.

An indication of the possible outcome of the deal was the approval by the federal government in October 1981 of the British sale of the sophisticated FH-70 towed howitzer to Saudi Arabia; this howitzer is jointly produced by the UK, FR Germany and Italy [12]. Several French–West German weapon systems, such as HOT, MILAN and Roland missiles and

Alpha Jet trainer/strike aircraft, are currently being sold world-wide under French arms export laws.

In recent years, Italy has emerged as the world's fourth largest exporter of major weapon systems after the United States, the Soviet Union and France. This boom is not entirely due to the quality of Italian weaponry. The export surveillance scheme enables firms to export to virtually any country in the world. The lack of government control over Italy's arms manufacturers is one of the aspects of the trade most strongly criticized inside and outside the country. The weapons exported include indigenously designed light warships, such as Lupo-class frigates, corvettes and fast patrol boats; missiles and aircraft as well as licence-produced helicopters of US design; and armoured vehicles from the USA and FR Germany. Italy's arms transfers are almost exclusively to Third World countries, with Libya as the single largest recipient.

The Soviet Union

The Soviet Union was the world's largest supplier of major weapon systems during the period 1979–81 (figure 3.1). However, the USSR still has a smaller number of customers than the USA, and it is less willing than the USA to allow licensed production of their major weapons. According to the SIPRI arms trade registers covering major weapons on order or in the process of delivery during 1981, the Soviet Union has current arms deals with 28 countries, while the corresponding figure for the United States is 67 countries. Furthermore, the registers identify 61 US major weapons being produced under licence outside the USA, while the Soviet Union has only 10 similar arrangements: these are with Czechoslovakia, Poland and India.

Soviet arms exports are otherwise guided by the same political and economic motives as those of the United States. Arms transfers serve as a means of establishing a presence in regions important to the Soviet Union or to counter Western interests. Military sales and assistance often provide the opening wedge for a variety of other contacts which would otherwise have been difficult to achieve. An arms agreement with a developing country has been the point of departure for most Soviet advances in the Third World, beginning with the first Soviet arms deal negotiated with Egypt in 1955–56.

One attractive feature of Soviet military assistance from a Third World point of view has traditionally been low prices and favourable credit terms. The prices charged have naturally varied with the type and quality of the equipment, but Soviet prices have on the whole been lower than Western prices for comparable equipment. Credits have usually been made available at a 2 per cent interest rate and a 10-year credit period [13]. This situation has been changing during the past couple of years.

> The USSR has recently faced some difficulties in sustaining such terms for military aid because of its declining economy; has increasingly had to seek hard currency payments for its military equipment; and since 1977, has often required a substantial cash down payment. In the case of recent jet fighter sales to Zambia, it offered only seven years credit at commercial rates. [14a]

As with Western arms suppliers, the Soviet Union needs arms exports as a way of lowering domestic procurement costs. Sales for hard currency have almost entirely supplanted the favourable terms of earlier years, especially when the clients are oil-producing countries such as Algeria, Iraq or Libya. One result of this may be the diversion of domestic stocks for export purposes. It is believed that the surprisingly slow introduction into service of the T-72 main battle tank in the WTO armies is partly explained by large exports to Libya, Syria and other oil-producing countries in the Middle East and North Africa.

Officially, the USSR refused to supply either side in the Iran–Iraq war during 1981. In spite of this, Soviet equipment found its way to both antagonists through countries allied to the Soviet Union. Poland delivered more than 100 (some sources report 300) T-55 tanks to Iraq, while North Korea reportedly shipped Soviet weapons to Iran [14b]. It is also possible that the Israeli air raid on the nuclear reactor in Baghdad made the USSR

Table 3.3. Rank order of the 20 largest Third World major-weapon importing countries, 1979–81

Percentages are based on SIPRI trend indicator values, as expressed in US $ million, at constant (1975) prices.

Importing country	Percentage of total Third World imports	Importing country	Percentage of total Third World imports
1. Libya	*9.0*	11. Peru	*2.7*
2. Saudi Arabia	*8.9*	12. Algeria	*2.6*
3. Iraq	*7.7*	13. South Korea	*2.5*
4. Syria	*7.3*	14. Argentina	*2.2*
5. Israel	*6.8*	15. Indonesia	*2.0*
6. India	*5.1*	16. Cuba	*1.7*
7. South Yemen	*3.9*	17. Thailand	*1.6*
8. Egypt	*3.9*	18. Chile	*1.6*
9. Viet Nam	*3.7*	19. Kuwait	*1.6*
10. Morocco	*2.8*	20. Taiwan	*1.5*
		Others	*20.9*
		Total	*100.0*
		Total value[a]	**25 971**

[a] Values include licence production.

Source: SIPRI data base.

resume direct deliveries of spare parts and arms to Iraq, although this has not been confirmed. Kuwait and Jordan, the latter a traditional client for US weapons, are two other Middle East countries opting for Soviet military equipment, mainly surface-to-air missiles.

In South Asia, it is likely that the US decision to sell F-16 fighters to Pakistan will trigger new arms deals between India and the Soviet Union —deals that will be in excess of the $1.6 billion arms credit package concluded between the two countries in 1980. In Afghanistan, the USSR has introduced MiG fighters, Mi-24 Hind helicopter gunships and numerous infantry fighting vehicles in the war against the resistance, but no major weapons are being transferred to Afghan government forces, which are apparently regarded as unreliable. In Central America, Cuba received during 1981 MiG-21/23 fighters, T-62 tanks, a Koni-class frigate and other equipment.

The Soviet Union has, together with other major arms suppliers, been faced with the prospect that the recipients might use their weapons for purposes not congruent with the intentions of the supplier. However, the Soviet Union is using arms transfers as an important instrument for maintaining and expanding its influence in the Third World. Arms transfers play a far greater role than economic aid or trade in this respect; it is virtually the only area in which they have successfully rivalled the West.

Third World suppliers

Arms exporting countries in the Third World can be divided into two categories: those which export domestically produced weapons, whether indigenously designed or produced under licence (notably Brazil, Israel, South Africa, India and Argentina), and those which re-export arms originally purchased from the industrialized countries (for example Egypt, Libya and Saudi Arabia). The Third World share of the global trade in major conventional weapons is comparatively small, 2.4 per cent for the period 1979–81, but it is a growing share (see figure 3.1 and table 3.4). Third World countries also export large quantities of small arms. Third World arms producers sell arms mainly for economic reasons. Because of lower unit prices—made possible by lower production costs—it is above all other Third World countries that buy these weapons. Political preferences are of lesser importance: "We're looking to the Third World, and we'll sell to the right, the left and the center", says one Brazilian government arms sales director [15].

Brazil has a booming arms industry. The Engesa company reportedly sells approximately 1 000 armoured vehicles a year to 32 countries, mostly on arms-for-oil terms to OPEC members in Africa and the Middle East. Brazilian rifles and machine-guns are in service in Angola and Congo.

Table 3.4. Rank order of the six largest Third World major-weapon exporting countries, 1979–81

Percentages are based on SIPRI trend indicator values, as expressed in US $ million, at constant (1975) prices.

Exporting country	Percentage of total Third World export
1. Brazil	*45.6*
2. Israel	*21.1*
3. Libya	*12.3*
4. South Korea	*8.2*
5. Egypt	*6.2*
6. Saudi Arabia	*1.6*
Others	*5.0*
Total	*100.0*
Total value	**993**

Source: SIPRI data base.

The Avibras company sells, among other things, air-to-ground missiles to Iraq, and Embraer markets a wide range of aircraft including jet trainers, counter-insurgency aircraft and transports. In 1981 Brazil started deliveries of the Xingu trainer/light transport jet to the French Air Force.

In 1979 Israel reportedly sold arms of a total value of $600 million, a figure that rose to $1.2 billion in 1980 [16]. Israel produces the Kfir jet fighter, Shafrir and Gabriel missiles, the Merkava tank and Reshef missile boats, several of which have been sold to South Africa. It is, however, mainly through exports of defence electronics, small arms and ammunition that Israel has reached its position as one of the world's leading arms exporters. More than 300 000 rounds of 105-mm HEAT (high-explosive anti-tank) tank ammunition has been sold, including a $40 million deal concluded with Switzerland in September 1981. The Galil rifle is another prominent export item; 10 000 are now being supplied to the Guatemalan Army under an agreement worth $6 million [17]. The most conspicuous Israeli arms transfer during 1981 is the sale to Iran of ammunition, re-furbished jet engines, spare parts for US-built M-48 tanks and tyres for F-4 Phantom fighters. Some of these items were shipped from Tel Aviv to Teheran by a British private arms dealer in an Argentine aircraft via Larnaca Airport in Cyprus [18]. It is also, incidentally, via this airport that the French Mirage F-1 fighters are being ferried to Iraq.

Other Third World deliveries to Iran during 1981 include 190 Soviet-built T-54/55/62 tanks, artillery shells and more spares for the M-48s from Libya. Egypt has, on the other hand, provided Iraq with $25 million worth of military equipment delivered via Oman [19]. Other recipients of weapons from Egypt include Chad, Somalia, Sudan and the Afghan resistance. Sudan has also been receiving a number of old US tanks, probably M-41s and M-47s, from Saudi Arabia [20].

References

1. Kearns, G., 'CAT and dogma: the future of multilateral arms transfer restraint', *Arms Control*, Vol. 2, No. 1, May 1981, p. 7.
2. Presidential Directive on conventional arms transfer policy, The White House, Office of the Press Secretary, 9 July 1981.
3. *FY 1982 Security Assistance Authorization*, Hearing before the Committee on Foreign Relations, US Senate, 97th Congress, 1st Session (US Government Printing Office, Washington, D.C., 31 March 1981), pp. 14–15.
4. 'The US has lost a lot of years', interview with Egyptian Defence Minister Abu-Ghazala, *Armed Forces Journal International*, September 1981, p. 50.
5. Bernstein, C., 'US weapons for Afghanistan', *Chicago Tribune*, 23 July 1981.
6. *Financial Times*, 22 June 1981.
7. *The Times*, 14 April 1981.
8. *Financial Times*, 5 March 1981.
9. *CAAT Newsletter*, No. 46, 11 February 1981.
10. *The Times*, 6 October 1981.
11. Philipp, U., 'German arms exports—the debate warms up', *International Defence Review*, April 1981, p. 417.
12. *Frankfurter Allgemeine Zeitung*, 13 October 1981.
13. Pajak, R., 'Soviet arms transfers as an instrument of influence', *Survival*, No. 4, 1981 (IISS, London), p. 167.
14. Cordesman, A., 'US and Soviet competition in arms exports and military assistance', *Armed Forces Journal International*, August 1981.
    (a) —, p. 67.
    (b) —, p. 68.
15. *Newsweek*, 9 November 1981, p. 34.
16. *Financial Times*, 12 November 1981.
17. *Baltimore Sun*, 4 February 1981.
18. *Sunday Times*, 26 July 1981.
19. *Milavnews*, May 1981, p. 17.
20. *Defence & Foreign Affairs Daily*, 20 May 1981.

4. Strategic nuclear weapons

Square-bracketed numbers, thus [1], *refer to the list of references on page* 91.

I. Introduction

The balance between the two great powers in intercontinental nuclear weapons is becoming increasingly unstable. The number of warheads has multiplied, they have been made much more accurate, and many of them are targeted on the silos of the other side. Each side is claiming that the other side is trying for some kind of first-strike capability, while denying that its own objective is anything but defensive. Thus in August 1980 the then US Secretary of Defense, Harold Brown, said:

> In the future, Soviet military programs could, at least potentially, threaten the survivability of each component of our strategic forces. For our ICBMs, that potential has been realized, or close to it. The Soviets are now deploying thousands of ICBM warheads accurate enough to threaten our fixed Minuteman silos. For our bombers the threats are more remote, and for SLBMs, more hypothetical. But the Soviets are developing, for employment in the mid-1980s, airborne radars and anti-aircraft missiles to shoot down our penetrating B-52s. And they are searching intensively for systems to detect and destroy our ballistic missile submarines at sea. These Soviet efforts cannot be ignored. [1]

On the Soviet Union's side, a recent publication of the Ministry of Defence says:

> The M-X, now in its final stage of development, is designed as a first-strike weapon. . . . According to the tactical and technical specifications of the US Defense Department, the [Trident II] missile will have practically the same combat capability as the M-X ICBM, that is, it will be a first-strike weapon. . . . The agreed schedule of the Pentagon plans for building up strategic offensive armaments and deploying anti-missile and space defensive systems is timed to complete the development of a so-called first-strike potential in the 1980s. [2]

Given that both sides have substantial numbers of submarine-launched ballistic missiles (SLBMs), and given that the threat of a first strike against these is, as Harold Brown says, hypothetical, this stress on the present or potential first-strike capability of the other side is at first sight puzzling. The scenario suggested on the United States side goes like this. The Soviet Union launches a strike which eliminates all US land-based missiles. It still has enough strategic nuclear weapons in reserve to inhibit the United States from making any reply—since that reply would then bring total devastation on the United States. So there is no US retaliation, either with submarine-launched missiles or with the cruise missiles of the bomber fleet.

There are a number of implausibilities in this scenario, discussed below. Nonetheless it is used, on the United States side, as justification for their new strategic weapon plans—including the development of missiles such as the MX and Trident II, with much greater accuracy than the missiles they replace; the search for a less vulnerable basing system for the MX missile; and the multiplication of cruise missiles. There is also renewed discussion in US strategic journals of the need to establish a 'launch-on-warning' system. To prevent US land-based missiles being caught in their silos, these missiles should themselves be fired as soon as there was evidence that Soviet missiles could reach their target. They should be fired on the basis of a computer analysis of the evidence from various detection devices, without reference to the President [3].

So now, instead of what might once have seemed to be a stable system of deterrence—a balance of mutually assured destruction (MAD)—we have the fear of a first strike being used as the rationale for the very big increases now in prospect in strategic weapon programmes and procurement. Between them, the two great powers, with the nuclear weapons at their command, have a total destructive power which is probably equivalent to about half a million Hiroshima bombs: but that is not enough. There can be no better example of the way in which developments in weapon technology—in this case the increasing accuracy of intercontinental ballistic missiles (ICBMs)—lead to a reduction rather than to an increase in security.

Reality or myth?

Can one conceive of a Soviet leadership, or a US President, ordering an attempt at a first strike—except as a pre-emptive move, in the belief that the other side was about to do the same? It is not legitimate simply to deduce from the increasing accuracy of ICBMs that governments are seriously considering the possibility of launching a first strike in cold blood. The constraints that inhibit any leadership from considering such an option were set out in a classic statement by Henry Kissinger, then Secretary of State:

> Indeed neither side has even tested the launching of more than a few missiles at a time; neither side has ever fired in a North–South direction as they would have to do in wartime. Yet initiation of an all-out surprise attack would depend on substantial confidence that thousands of re-entry vehicles launched in carefully coordinated attacks ... would knock out all their targets thousands of miles away with a timing and reliability exactly as predicted, before the other side launches any forces to pre-empt or retaliate, and with such effectiveness that retaliation would not produce

unacceptable damage. Any miscalculation or technical failure would mean national catastrophe. Assertions that one side is 'ahead' by the margins now under discussion pale in significance when an attack would depend on decisions based on such massive uncertainties and risks. [4]

There is the uncertainty indicated by the increase of accuracy. The 'circular error probability' (CEP)—the indicator normally used—gives the radius of the circle within which half the missiles will fall; half, it must not be forgotten, will fall outside it. The measure of accuracy has, of course, been calculated on trajectories different from those which would actually be employed in an attack. There is the uncertainty of the missile's reliability—not all test firings by any means have been successful. There is the problem in any attempt to compensate for these uncertainties by firing more than one warhead at each silo—the problem of fratricide. It is highly likely that the first of a series of warheads to explode will impair the function of other first-strike warheads in the vicinity before they, in their turn, can explode. Between the shock wave from the first explosion and the development of its mushroom cloud, there is apparently a 'window' where a second warhead may get through, and the duration of the window can be estimated. But remarkable co-ordination would be needed if such windows were to be hit 1 000 times, and the proof of performance would require atmospheric testing in a manner which cannot now be attempted.[1]

Even if one were to suppose that a first strike successfully eliminated virtually all land-based missiles, the leadership which ordered it could not possibly be confident that there would be no retaliation with the submarine-launched ballistic missiles remaining. It has been estimated that an attempt at a first strike on US land-based missiles would produce a quantity of radioactive dust which (according to the Office of Technology Assessment) would kill within a month between 2 and 20 million Americans [5]. What confidence would the Soviet leadership have that under such circumstances the US President would decline to order any retaliation from the submarine-launched ballistic missiles or the bombers with cruise missiles which were still available to him?

The arguments, of course, also apply to an attempted US first strike against Soviet land-based missiles.

To attempt the launch of a first strike against just one part of another country's strategic weaponry would be an act with a very great risk of total catastrophe to the power which launched it. As a realistic technological and political option it lies in the realm of myth. Yet it is this theme —the fear of a first strike—which is presented as the justification for the major new developments in strategic nuclear weaponry which seem likely

[1] The statistical uncertainties of a first strike are discussed in reference [5].

to come about in the next decade. Unfortunately myths are often powerful in political affairs.

The sections which follow discuss, first, the developments in Soviet strategic nuclear forces, and second, the developments on the US side. The section on Soviet strategic nuclear forces is mainly about installations which already exist. Information about Soviet future trends in this matter is, as usual, scanty. On the US side, on the other hand, there is a great deal of information now about future proposals and plans, and these plans are an important part of the story. The Soviet section therefore is mainly about things which the Soviet Union has already done. The United States section is mainly about things which are planned for the future.

II. Developments in Soviet strategic nuclear weapons

The current stock of Soviet intercontinental strategic nuclear delivery systems consists of 1 398 ICBM launchers, 950 SLBM launchers and 156 long-range bombers. Between them, these delivery systems are loaded with about 7 000 nuclear warheads—a number which will probably increase over the next few years. Compared with the United States, a larger proportion of total warheads are deployed on land-based ICBMs and a smaller proportion on submarines. The previous section puts forward arguments for suggesting that land-based ICBMs have not, in any realistic sense, become vulnerable to total elimination in a first strike. Insofar as governments nevertheless may have come to believe that is the case, then the situation may appear to be more serious for the Soviet Union than for the United States. However, the Soviet Union is concentrating on increasing the capability and invulnerability of its submarine-based strategic forces; this may change the situation before the end of the decade.

The sections which follow discuss developments in each of the three categories—land-based missiles, submarine-launched missiles and bombers; and there are some comments on the Soviet system of air defence.

Land-based missiles

In the second half of the 1960s, the Soviet Union's strategic weapon programme concentrated on increasing the number of land-based launchers to some kind of parity with the United States. The main developments of the 1970s have consisted not so much of further increases in the number of launchers but rather of replacement of old missiles by more modern ones—by SS-17s, SS-18s and SS-19s. Over half the total number of land-based launchers are now in these new categories. The total stock consists now of 520 SS-11s, 60 SS-13s, 150 SS-17s, 308 SS-18s and about 360 SS-19s. The great majority of the SS-17s, -18s and -19s are equipped with MIRVs (multiple independently targeted re-entry vehicles).

The SS-17 has been deployed in converted SS-11 silos and is the least accurate of the newer Soviet ICBMs. It is believed that it can deliver four warheads over a range up to 10 000 km and is 'cold-launched'. This technique minimizes launch damage to the silo, which can then be reloaded with a missile after the first one has been fired. It would probably take a few days to make a silo ready for refiring a missile—therefore this facility does not violate the provisions of the SALT II Treaty which preclude a rapid reloading capability for ICBM launchers.

The SS-18 is the largest Soviet ICBM, twice as large as the proposed US MX missile. It is believed to be capable of delivering 8 or 10 MIRVed warheads over a range of up to 10 000 km. (If the range decreases, the number of warheads can of course be increased.) The SS-18 is also cold-launched. Together with the SS-19, it is judged to be the most accurate of the Soviet ICBMs.

The SS-19 is comparable in size to the proposed US MX missile. It is believed that it can deliver 6 warheads to a range of 9 000 km, and uses a hot-launch technique in which the missile's engine is ignited while the missile is in its silo.

The replacement of old missiles by these newer types will probably be complete by the mid-1980s. It is also anticipated that the Soviet Union will develop solid-propellant ICBMs to supplement or replace some of the current liquid-propellant systems. Solid-propellant ICBM development and deployment could give the Soviet Union additional flexibility in handling and in basing their missile forces.

Sea-based strategic weapons

The Soviet Union has since the 1960s developed a series of some 60 modern nuclear ballistic missile submarines. The trend has been to increase the number of missile tubes; the range of the missiles has also been steadily extended. The Hotel-class of the early 1960s was followed by the Yankee-class with 12–16 tubes and with missiles which had a range of 3 000–4 000 km. The most modern class of Soviet ballistic missile submarines which is operational is the Delta-class, also with 12–16 missile tubes, but with missiles which have a range of 8 000–9 000 km. The SS-N-18 missile which is carried by the Delta III submarine is a liquid-propelled, two-stage missile; it was the first Soviet submarine-launched ballistic missile to carry multiple warheads. These missiles can be fired at most targets in the USA from Soviet home waters. The submarines equipped with such long-range missiles therefore do not have to expose themselves to any important extent to US anti-submarine warfare systems.

In 1980, the Soviet Union launched a new and very large strategic nuclear submarine, the Typhoon. This is believed to be about 160 m long, to displace about 25 000 tons submerged, and to carry 20 ballistic missiles. This class of submarine should become operational from the mid-1980s onwards and be equipped with a new, more accurate ballistic missile, the SS-NX-20. This missile will probably have up to 12 warheads, and will also be able to cover most targets in the United States from Soviet home waters (its range is 4 200 nautical miles). It is suggested that the Typhoon may be deployed under the ice of the Arctic Ocean, as further protection against US anti-submarine tactics.

Bombers

The Soviet Union has not done much to modernize its long-range bomber fleet. Its heavy bomber capability continues to rest principally on the small and ageing 'Bison–Bear' force, consisting of some 100 turbo-prop Bears and 56 Bisons. Both these types were first deployed in the mid-1950s. There have from time to time been reports of a new heavy bomber. In the US Department of Defense annual report for the fiscal year 1979, Secretary Harold Brown stated: "We now expect to see the first prototype of a new heavy bomber in the near future." It was then expected to fly during 1979. There was a similar report, in December 1981, of a variable-geometry swept-wing bomber photographed on the apron at the Ramenskoye flight test centre [6]. It is obviously too early to say whether there will or will not be any substantial production of a new Soviet heavy bomber.

The Soviet air defence system

Unlike the United States, the Soviet Union maintains a formidable air defence system. This consists of a large number of air defence interceptor aircraft and a very large array of surface-to-air missiles. Now that the United States is planning to deploy a large number of cruise missiles, the Soviet Union will probably upgrade its defence systems specifically to deal with the cruise missile threat. This will probably involve the development of a more effective Airborne Warning and Control System (AWACS) to detect low-altitude penetrators.

One problem for the Soviet Union here is that it is almost certainly more expensive to deploy an effective defence system against cruise missiles than to deploy the cruise missiles themselves. Some experts argue that the main reason why the United States is proposing to deploy

Table 4.1. Soviet strategic weapon delivery capability (mid-1982)

Vehicle	Number of delivery vehicles deployed	Number of warheads per delivery vehicle	Total delivery capability (number of warheads)	Total yield per delivery vehicle (Mt)	Total delivery capability (Mt)
MIRVed vehicles					
SS-17	150	4	600	2	300
SS-18	308	8	2 464	5	1 540
SS-19	360	6	2 160	3	1 080
SS-N-18[a]	256	3	768	0.6	154
Sub-total	1 074		5 992		3 074
Non-MIRVed vehicles					
'Bison' (bombs)	56	2	112	2	112
'Bear' (bombs)	100	3	300	3	300
SS-11	230	1	230	1	230
SS-11 (MRV)	290	3	870	0.6	174
SS-13	60	1	60	1	60
SS-N-5[a]	18	1	18	1	18
SS-N-6[a]	102	1	102	1	102
SS-N-6[a] (MRV)	272	3	816	0.6	163
SS-NX-17[a]	12	1	12	1	12
SS-N-8[a]	290	1	290	1	290
Sub-total	1 430		2 810		1 461
Total	**2 504**		**8 802**[b]		**4 535**

[a] SLBM.
[b] ICBMs carry 72 per cent of the total number of warheads, SLBMs 23 per cent and bombers 5 per cent.

Table 4.2. US strategic weapon delivery capability (mid-1982)

Vehicle	Number of delivery vehicles deployed	Number of warheads per delivery vehicle	Total delivery capability (number of warheads)	Total yield per delivery vehicle (Mt)	Total delivery capability (Mt)
MIRVed vehicles					
Minuteman III	350	3	1 050	0.51	179
Minuteman III (Mk 12A)	200	3	600	1.05	210
Poseidon C-3[a]	320	10[b]	3 200	0.4	128
Trident C-4[a, d]	200	8	1 600	0.8	160
Sub-total	1 070		6 450		677
Non-MIRVed vehicles					
B-52 (SRAMS + bombs)	150[c]	12[d]	1 800	5.6	840
B-52 (bombs)	197[c]	4[d]	788	4	788
Titan II	52	1	52	9	468
Minuteman II	450	1	450	1.5	675
Sub-total	849		3 090		2 771
Total	**1 919**		**9 540**[e]		**3 448**

[a] SLBM.
[b] Average figure.
[c] Including heavy bombers in storage, etc., there are 573 strategic bombers.
[d] Operational loading. Maximum loading per aircraft may be eleven bombs, each of about one megaton.
[e] SLBMs carry 50 per cent of the total number of warheads, bombers 27 per cent and ICBMs 23 per cent.

so many cruise missiles is to try to force the Soviet Union into spending large sums on defensive measures. These would include AWACS aircraft constantly patrolling the Soviet borders to detect enemy cruise missiles, to alert and control interceptor aircraft and surface-to-air missiles and to destroy any incoming cruise missiles. In addition, ground radars would be used to detect low-flying missiles.

III. Developments in US strategic nuclear weapons

The main subject of interest on the US side is, of course, the current proposals for modernizing and expanding the US nuclear strategic armoury. There is clearly a wide gap between the Soviet and the US perceptions of the balance in strategic nuclear weapons. The Soviet perception is probably that they have at last achieved a rough parity in strategic nuclear weapons, and they signed the SALT II agreement as a document which gave expression to that parity. The US Administration has been persuaded that the Soviet Union is, in some sense, ahead in intercontinental nuclear weapons—particularly in its alleged ability to eliminate US land-based intercontinental missiles in a first strike. So the SALT II Treaty has not been ratified, and the new Administration is committed to attempt a radical revision of that Treaty; in the meantime it proposes very substantial new expenditure on strategic weapons. It is, of course, never certain that long-term plans of this kind will be fulfilled in their entirety: there may be economic, environmental or other constraints.

The US proposals cover the whole field of strategic nuclear weaponry —land-based missiles, strategic bombers, cruise missiles, submarine-launched ballistic missiles and command, control and communications. Each of these is discussed in turn.

Land-based intercontinental missiles

The United States will press ahead with the production of the new MX missile, whose characteristics are set out in table 4.3. At least 100 of these missiles are to be deployed, and they should be available by 1986. The proposal of the previous Administration, to put these missiles in multiple protective shelters, has now been rejected. This mobile base scheme involved shuttling 200 MX missiles between 4 600 horizontal shelters. The Reagan Administration's objection to this multiple shelter plan, apart from costs and public protests, was explained by Richard Perle, Assistant Secretary of Defense for International Security Policy, as follows: "The 4 600-shelter program was not persuasive; the USSR could have overcome it relatively easily without new technology in accuracy. They simply would increase the number of re-entry vehicles to go beyond 4 600 to 5 500 or to 6 600 to overcome it" [7]. Perle is assuming here that the Soviet Union would not be constrained by the provisions of the SALT II Treaty; these provisions would prevent the Soviet Union from increasing the numbers of its MIRVs in this way.

Table 4.3. Characteristics of the MX ICBM, compared with the Minuteman III

	MX	Minuteman III
Length (m)	21.5	18.2
Diameter (m)	2.3	1.8
Stages	3[a]	3
Weight (kg)	87 270	35 409
Propellant	Solid[a]	Solid
Guidance	Inertial	Inertial
Launching mode	Cold	Hot
Throw-weight (kg)	3 570	1 000
Range (km)	11 000	9 000
Number of MIRVs	10[b]	3

[a] The post-boost vehicle, which manoeuvres to guide the individual warheads after the three main stages burn out, is liquid-fuelled.
[b] This is the SALT II limit. However, the MX is designed to carry 11 Advanced Ballistic Re-entry Vehicles (about 500 kt each) or 12 ML 12A warheads (about 335 kt each).

Source: Congressional Research Service, Issue Brief Number IB77080.

The Administration's proposal was to put 35–40 of these MX missiles, as an interim measure, into existing ICBMs silos which would be further hardened. US silos are on average hardened to withstand an over-pressure of 2 000 psi (pounds per square inch). The proposal was to harden some 35–40 of them to 5 000 psi. In the view of the Secretary of Defense, these hardened silos should be able to withstand a Soviet attack from the time the missiles roll off the production line in 1985–86 until about 1987–88. This plan may, however, have been shelved.

The research and development on long-term basing options for this MX missile is to concentrate on three possibilities: continuously airborne patrol aircraft; deep underground basing; and ballistic missile defence. The first of these alternative modes envisages the use of an aircraft which is capable of flying for long periods of time over oceans, each carrying an MX missile for airborne launch. Deep underground silo basing would place the MX missiles in holes up to 1 000 m deep. The third option of developing ballistic missile defence would, of course, require the revision, or indeed the abandonment, of the Anti-Ballistic Missile Treaty. Secretary of Defense Caspar Weinberger has said, of the study of the possibilities of anti-ballistic missile defence: "If we find at the conclusion of the study that there is a far more effective system that would require revision of the treaty, I think it's fair to say we wouldn't hesitate to seek those revisions..." [8]. It does appear that, in more than one respect, the present US Administration is envisaging a future in which the development of strategic nuclear weaponry is not constrained by treaty. The decision among these various options—and the Secretary of Defense indicated that it is likely that more than one basing system would be recommended —should be made in time for the fiscal year 1984 budget.

Bombers

The bomber programme is the largest element in the total strategic programme costs (table 4.4). There are three main elements. The first, most immediate development is to upgrade the B-52Gs and B-52Hs so that they can carry some 3 000 cruise missiles; this deployment will begin this year. (The cruise missiles themselves are discussed in the next sub-section.) Secondly, 100 B-1B bombers will be built. The first bombers should be operational in 1986, and a fleet of some 90 aircraft by 1988 or 1989. The B-1B will be equipped to carry air-launched cruise missiles, probably 30 per aircraft. The B-1 bomber programme had been cancelled

Table 4.4. The US strategic weapon programmes for fiscal years 1982–87

Figures are estimates as of October 1981, and are in billions of US dollars, at FY 1982 prices.

Programme	Programme cost	Per cent of total
Bomber programme	63	*35*
B-52 upgrading, 100 B-1 bombers, cruise missiles, development of Stealth bomber		
Sea-based programmes	42	*23*
Trident submarines, Trident-2 missiles, cruise missiles		
Land-based programmes	34	*19*
MX missiles, hardening of silos, new basing system		
Strategic defence	23	*13*
6 AWACS aircraft, 5 squadrons of F-15s, R&D on anti-ballistic missiles		
Command and control systems	18	*10*
Satellites, communications to strategic weapon systems, hardening		
Total	**180**	**100**

Source: Defense Daily, 5 October 1981.

by President Carter in 1977 after four aircraft had been built. The argument of the previous Administration was that they could rely on the B-52s throughout the 1980s, and that they would develop the Advanced Technology Bomber for the 1990s. The new Administration considered that this represented a willingness to accept risks associated with an ageing and vulnerable B-52 force, and risks associated with the uncertain schedule and unproven capabilities of the Advanced Technology Bomber. Hence the decision to procure a fleet of 100 new bombers for the second half of the 1980s.

The third part of the bomber programme is the continuation of an intensive research and development programme for the Advanced Technology Bomber (the so-called 'Stealth' aircraft). By incorporating a number of technological developments, the bomber is to be given a very small radar cross-section to enable it to penetrate enemy air defence systems with a much reduced risk of detection. The research and development programme—which will be undertaken by a Northrop Corporation team—will also involve the development and fabrication of radar-absorbing materials to reduce radar detectability, aerodynamic and flight control systems with low observable characteristics and stealthy terrain-following and avoidance systems.

Cruise missiles

The new plan also accelerates production of the Boeing air-launched cruise missile (ALCM). The ALCM is a small, long-range, subsonic, nuclear-armed, winged vehicle to be deployed on B-52 strategic bombers and eventually on B-1B bombers. The missile is about 6 m long, weighing less than 1 360 kg, has a range of about 2 500 km and will be armed with a nuclear warhead with a yield of about 200 kt.

The ALCM could be launched by B-52s flying outside Soviet territory against air defence systems, to destroy their radars and ground-to-air missiles. Other B-52s would then be able to penetrate into Soviet territory to attack targets with their ALCMs. These missiles are accurate and can therefore be used against small, hardened military targets. They also have relatively small radar cross-sections and are therefore difficult to detect.

The first ALCM was produced in November 1981. The production rate was expected to be 7 per month by January 1982, rising to 14 per month at the end of the year. Twenty ALCMs would be deployed on each of 151 B-52Gs and B-52Hs. ALCMs will about double the number of nuclear weapons the US strategic bomber force carries.

The new Administration also plans to deploy Tomahawk cruise missiles —some of which will be nuclear-armed—on submarines and surface ships. These missiles are the same as the ground-launched missiles. The US

Navy plans to procure 1 720 Tomahawks between FY 1983 and FY 1987. Some of these will be mounted on launchers attached between the external and the pressure hulls of submarines. Each submarine will carry 12 such missiles. Others will be deployed on surface ships. The targets envisaged are land targets, as well as enemy ships. These cruise missiles may be equipped with either conventional or nuclear warheads; their deployment will therefore complicate the negotiation of any future strategic arms control treaties.

Sea-based strategic weapons

The first of the new Trident ballistic missile submarines, the *Ohio*, was commissioned in November 1981. It is approximately twice as large as a Poseidon-Polaris missile submarine: it will carry 24 Trident missiles, with a range of 7 500 km and eight 100-kt MIRVs. (This compares with 16 missiles on Poseidon submarines, with a range of 4 500 km and ten 40-kt MIRVs.) Eight such submarines are now being built; of these, five should be operational by 1987. The missile with which they are now fitted, the Trident C-4 (or Trident I), is also being retro-fitted into Poseidon submarines. Four of these are already in service with the new missile; another six soon will be.

The new strategic plan calls for the development and deployment of the Lockheed Trident D-5 (or Trident II) missile. This new SLBM is planned to have a range of 11 000 km, and carry up to 14 warheads each with a yield of 150 kt. These missiles will be much more accurate than those they replace. The submarines themselves will have more accurate navigation techniques, and the warheads may be fitted with terminal guidance, in which a radar device or laser will search the area around the target after the warhead has re-entered the atmosphere and guide the warhead very accurately to its target. Missiles like the Trident II will then be as accurate as land-based ICBMs, and are seen by the other side as first-strike weapons, capable of destroying enemy ICBMs in their hardened silos.

Strategic defence and communication and control systems

In strategic defence, the programme calls for a substantial upgrading of the North American air surveillance network, the replacement of five squadrons of F-106s with F-15s, the procurement of six additional AWACS surveillance aircraft and the pursuit of an operational anti-satellite system. There will also be substantial research and development in ballistic missile defence, including technology for space-based missile defence.

The communications and control systems programme has four main areas of expenditure. The first is improvement of the survivability, performance and coverage of radars and satellites which provide warning of a Soviet missile attack. The second is an upgrading of the capability of command centres, including in particular mobile command centres that could survive an initial attack. The third is the improvement of communications between the command centres and the strategic weapon systems themselves—ensuring two-way communications in many instances. The fourth is an R&D programme leading to a communications and control system that would survive the first nuclear attack. This probably refers in particular to hardening against the effects of the short but very powerful pulse of electromagnetic radiation (EMP) given off by a high-altitude nuclear explosion: this can produce a surge of voltage in electronic equipment large enough to damage it permanently.

References

1. Brown, H., Speech at the Naval War College, New Port, Rhode Island, 20 August 1980.
2. Soviet Ministry of Defence, *Whence the Threat to Peace* (Moscow, 1982), pp. 30, 34, 36.
3. Blair, S., 'MX and the counterforce problem: a case for silo development', *Strategic Review*, Summer 1981.
4. Kissinger, H., Speech on 22 March 1974, reprinted in *International Security*, Vol. 1, No. 1, Summer 1976, p. 187.
5. Anderson, J. E., 'First strike: myth or reality', *Bulletin of the Atomic Scientists*, November 1981.
6. *Aviation Week & Space Technology*, 14 December 1981, p. 17.
7. Perle, R., quoted in *Aviation Week & Space Technology*, 12 October 1981, p. 23.
8. Weinberger, C., in evidence before the US House Armed Services Committee, 6 October 1981, reported in *Defense Daily*, 7 October 1981.

Appendix 4A

US and Soviet strategic nuclear forces, 1973–82

Figures for 1973–76 are as of 30 June; figures for 1977–82 are as of 30 September.
The sources and notes follow the table.

		First in service	Range (nm)	Payload	1973	1974	1975	1976	1977	1978	1979	1980	1981	1982
Delivery vehicles														
Strategic bombers														
USA	B-52 C/D/E/F	1956	10 000	27 000 kg	149	116	99	83	83	83	83	83	83	83
	B-52 G/H	1959	10 900	34 000 kg	281	274	270	265	265	265	265	265	265	264
	(FB-111	1970	3 300	17 000 kg	66	66	66	66	66	66	66	65	64	63)
USSR	Mya-4 'Bison'	1955	5 300	9 000 kg	56	56	56	56	56	56	56	56	56	56
	Tu-95 'Bear'	1956	6 800	18 000 kg	100	100	100	100	100	100	100	100	100	100
	(Tu-22M 'Backfire'	1975	4 000	9 000 kg	–	–	–	12	24	36	48	60	72	84)
				Long-range bomber total: USA	**430**	**390**	**369**	**348**	**348**	**348**	**348**	**348**	**348**	**347**
				USSR	**156**	**156**	**156**	**156**	**156**	**156**	**156**	**156**	**156**	**156**
Submarines, ballistic missile-equipped, nuclear-powered (SSBNs)														
USA	With Polaris A-2	1962	n.a.	16×A-2	8	6	3	–	–	–	–	–	–	–
	With Polaris A-3	1964	n.a.	16×A-3	13	13	13	13	11	10	10	5	5	–
	With Poseidon C-3 conv.	1970	n.a.	16×C-3	20	22	25	28	30	31	31	25	20	20
	With Trident C-4 conv.	1979	n.a.	16×C-4	–	–	–	–	–	–	–	6	11	11
	With Trident C-4	1980	n.a.	24×C-4	–	–	–	–	–	–	–	–	1	1
USSR	'Hotel II' conv.	1963	n.a.	3×'SS-N-5'	7	7	7	7	7	7	7	6	6	6
	'Hotel III' conv.	1967	n.a.	6×'SS-N-6'	1	1	1	1	1	1	1	1	1	1
	'Yankee'	1968	n.a.	16×'SS-N-6'	33	33	33	33	33	33	33	29	27	23
	'Yankee II'	1974	n.a.	12×'SS-NX-17'	–	1	1	1	1	1	1	1	1	1
	'Golf IV' conv.	1972	n.a.	4×'SS-N-8'	1	1	1	1	1	1	1	1	1	1
	'Hotel IV' conv.	1972	n.a.	6×'SS-N-8'	1	1	1	1	1	1	1	1	1	1
	'Delta I'	1973	n.a.	12×'SS-N-8'	1	7	12	18	18	18	18	18	18	18
	'Delta II'	1977	n.a.	16×'SS-N-8'	–	–	–	–	4	4	4	4	4	4
	'Delta III'	1978	n.a.	16×'SS-N-18'	–	–	–	–	–	2	4	10	12	16
				Submarine total: USA	**41**	**41**	**41**	**41**	**41**	**41**	**41**	**36**	**37**	**32**
				USSR	**44**	**51**	**56**	**62**	**66**	**68**	**70**	**71**	**71**	**71**
				Modern subs: USSR	**34**	**41**	**46**	**52**	**56**	**58**	**60**	**62**	**62**	**62**
SLBM (Submarine-launched ballistic missile) launchers on SSBNs														
USA	Polaris A-2	1962	1 500	1×1 Mt	128	96	48	–	–	–	–	–	–	–
	Polaris A-3	1964	2 500	3×200 kt (MRV)	208	208	208	208	176	160	160	80	80	–

	Poseidon C-3	1970	2 500	10 × 40 kt (MIRV)	320	352	400	448	480	496	496	400	320	320
	Trident C-4	1979	4 000	8 × 100 kt (MIRV)	–	–	–	–	–	–	–	96	200	200
USSR	'SS-N-5'	1963	700	1 × 1 Mt	21	21	21	21	21	21	21	18	18	18
	'SS-N-6 mod. 1'	1968	1 300	1 × 1 Mt	534									
	'SS-N-6 mod. 2' conv.	1973	1 600	1 × 1 Mt	–	534	534	534	534	534	534	470	438	374
	'SS-N-6 mod. 3' conv.	1973	1 600	2 × 200 kt (MRV)	–									
	'SS-N-8'	1973	4 300	1 × 1 Mt	22	94	154	226	290	290	290	290	290	290
	'SS-NX-17'	n.a.	. .	1 × 1 Mt (MIRV-cap.)	–	12	12	12	12	12	12	12	12	12
	'SS-N-18'	n.a.	4 050	3 × 200 kt (MIRV)	–	–	–	–	–	32	64	160	192	256
				SLBM launcher total: USA	**656**	**656**	**656**	**656**	**656**	**656**	**656**	**576**	**600**	**520**
				USSR	**577**	**661**	**721**	**793**	**857**	**889**	**921**	**950**	**950**	**950**
ICBMs (*Intercontinental ballistic missiles*)														
USA	Titan II	1963	6 300	1 × 10 Mt	54	54	54	54	54	54	53	52	52	52
	Minuteman I	1963	6 500	1 × 1 Mt	190	100	–	–	–	–	–	–	–	–
	Minuteman II	1966	7 000	1 × 1.5 Mt	500	500	450	450	450	450	450	450	450	450
	Minuteman III conv.	1970	7 000	3 × 170 kt (MIRV)	310	400	550	550	550	550	550	550	450	350
	Minuteman III impr.	1979	7 000	3 × 350 kt (MIRV)	–	–	–	–	–	–	–	–	100	200
USSR	'SS-7 Saddler'	1962	6 000	1 × 5 Mt	190	190	190	130	30	2	–	–	–	–
	'SS-8 Sasin'	1963	6 000	1 × 5 Mt	19	19	19	19	19	–	–	–	–	–
	'SS-9 Scarp'	1966	6 500	1 × 10–20 Mt	288	288	288	248	188	128	68	–	–	–
	'SS-11 mod. 1'	1966	5 700	1 × 1 Mt										
	'SS-11 mod. 2' conv.	1973	. .	1 × 1 Mt										
	'SS-11 mod. 3' conv.	1973	. .	3 × 200 kt (MRV)	990	1 010	1 030	950	860	750	640	580	580	520
	'SS-11 mod. 3'	1973	. .	3 × 200 kt (MRV)										
	'SS-13 Savage'	1969	4 400	1 × 1 Mt	60	60	60	60	60	60	60	60	60	60
	'SS-18 mod. 1/mod. 3'	1976	5 500	1 × 10–20 Mt	–	–	–							
	'SS-18 mod. 2' conv.	1977	. .	8 × 500 kt (MIRV)	–	–	–	60	120	180	240	308	308	308
	'SS-19' conv.	1976	5 000	6 × 500 kt (MIRV)	–	–	–	80	120	180	240	300	300	360
	'SS-17' conv.	1977	. .	4 × 500 kt (MIRV)	–	–	–	–	50	100	150	150	150	150
				ICBM total: USA	**1 054**	**1 054**	**1 054**	**1 054**	**1 054**	**1 054**	**1 053**	**1 052**	**1 052**	**1 052**
				USSR	**1 547**	**1 567**	**1 587**	**1 547**	**1 447**	**1 400**	**1 398**	**1 398**	**1 398**	**1 398**
				Total, long-range bombers and missiles: USA	**2 140**	**2 100**	**2 079**	**2 058**	**2 058**	**2 058**	**2 057**	**1 976**	**2 000**	**1 919**
				USSR	**2 280**	**2 384**	**2 464**	**2 496**	**2 460**	**2 445**	**2 475**	**2 504**	**2 504**	**2 504**
Nuclear warheads														
Independently targetable warheads on missiles: USA					5 210	5 678	6 410	6 842	7 130	7 274	7 273	7 000	7 032	. .
USSR					2 124	2 228	2 308	3 160	3 894	4 393	4 937	5 920	6 848	. .
Total warheads on bombers and missiles, official US estimates: USA					6 784	7 650	8 500	8 400	8 500	9 000	9 200*	9 200*	9 000*	. .
USSR					2 200	2 500	2 500	3 300	4 000	4 500	5 000*	6 000*	7 000*	. .

* 1 January.

Sources and notes for appendix 4A

Sources: The main sources and methodology of this appendix are described in the *SIPRI Yearbook 1974*, pp. 108–109, where a comparable table for the decade 1965–74 appears.

The earlier table has been updated on the basis of material published in the *Annual Report* of the US Secretary of Defense for the fiscal years 1976 to 1983 (US Government Printing Office, Washington, D.C., 1975–1982) and the statements on *US Military Posture* by the Chairman of the Joint Chiefs of Staff for the same eight years.

The version of this table for 1967–76 which appeared in the *SIPRI Yearbook 1976*, pp. 24–27, included revised estimates of the numbers of US strategic submarines and SLBMs of various types, based on the dates of overhaul and conversion of each submarine given in *Jane's Fighting Ships* (Macdonald & Co., London, annual), *Ships and Aircraft of the US Fleet* (Naval Institute Press, Annapolis, Maryland, recent editions), and US Senate Committee on Appropriations annual *Hearings* on naval appropriations. The revised series has been continued, based on the same sources.

The estimates of the numbers of US strategic bombers were revised in the table for 1968–77 which appeared in the *SIPRI Yearbook 1977*, pp. 24–28. The revised series, continued here, is based on a narrow definition of 'active aircraft'—the only definition which permits a consistent time series to be constructed from public data—taking the authorized 'unit equipment' (number of planes per squadron) of the authorized numbers of squadrons of each type of plane and adding a 10 per cent attrition and pipeline allowance (or lower when it is known that adequate numbers of spare aircraft are lacking).

A version of the table covering the period 1967–78 appeared in the brochure containing the SIPRI Statement on World Armaments and Disarmament, presented at the UN General Assembly Special Session devoted to Disarmament on 13 June 1978. That table listed three configurations of Soviet submarine, also shown here ('Hotel III', 'Yankee II' and 'Delta III'), which had not been previously reported. Reference to these configurations, as well as to the 'Hotel IV' and 'Golf IV' SS-N-8 test conversions, are given in the defence statements of the US Secretary of Defense and Joint Chiefs.

Notes:

Dates of deployment

The estimates for the year 1982 are planned or expected deployments.

In the case of the official US estimates of total warheads on bombers and missiles (the last two rows of the table), the estimates for 1979–81 refer to 1 January. All other estimates in the table follow the more usual practice of official US accounts—which are the main source of the data—by referring to the closing date of the US government fiscal year.

US SLBMs and submarines

The number of US submarines and the corresponding SLBMs are derived by treating all submarines under conversion as though they carry their former load until the conversion is completed (shipyard work finished), and they take on their new load from the date of completion. This method, the only exact procedure feasible with public data, differs from the practice in some official US accounts of excluding from the estimates of *total force loadings* (warheads on bombers and missiles) the loads that would be carried by submarines undergoing conversion and treating the submarines as under conversion until the date of their first subsequent operational deployment at sea.

The first of 12 Poseidon-equipped submarines which are to be backfitted with the Trident I (C-4) missile began conversion in the autumn of 1978 and became operational in October 1979. The first Trident submarine, with 24 launch tubes for the Trident I or Trident II missile (the latter now under development), began sea trials in 1981 and is therefore considered operational as of 31 September 1981.

The maximum payload of the Poseidon missile is 14 warheads, rather than the 10 shown in the table. It is estimated that, today, these missiles actually carry only 10 warheads each, an off-loading undertaken to compensate for poorer-than-expected performance by the missile propulsion system, so that the design range of 2 500 nautical miles can be reached. (In *Combat Fleets of the World 1978/79* (US Naval Institute Press, Annapolis, Maryland, 1978) Jean Labayle Couhat suggests that a range of 2 500 nautical miles can be reached with a 14-warhead payload and that reduction of the payload to 10 warheads increases the range to 3 200 nautical miles.) An article in the *New York Times* and an unofficial US Defense Department report, both from the autumn of 1980, have stated that, as the longer-range Trident missiles are phased in, covering more distant targets, the payload of the remaining 304 Poseidon missiles will revert to the originally designed 14 warheads. This will add a total of 1 216 warheads to the US SLBM force in the early 1980s.

US ICBMs

Three hundred of the 550 Minuteman III missiles are being backfitted with the Mark 12A re-entry vehicle, each of which will carry a 350-kt warhead. Moreover, NS-20 improvements in Minuteman III guidance have brought the expected accuracy (circular error probability) of this missile to about 190 m. This gives the current 170-kt Minuteman III warhead a better than 50:50 chance of destroying a Soviet missile silo hardened to 1 000–1 500 psi, and two such warheads in succession (barring 'fratricide' effects) about an 80 per cent probability of kill. The hard-silo kill probability of the new 350-kt warhead, given 190-m accuracy, will be about 57 per cent for one shot and close to 95 per cent for two shots.

MIRVed warheads on Soviet ICBMs

The original Soviet ICBM MIRVing programme is coming to an end, with a total of 818 ICBM silos converted to MIRV-capable launchers. The last of 308 SS-9 silos converted to hold the SS-18 were completed in 1980, and the 60 last SS-11 silos converted to hold the SS-19 are expected to be equipped with the SS-19 missile in 1982.

The exact numbers of MIRVed and unMIRVed versions of the SS-17, -18 and -19 are not known. All launchers for these missiles are counted as MIRV launchers for the purpose of the current understanding between the USA and the USSR to abide by the terms of the unratified SALT II Treaty.

Soviet and US bomber aircraft

The long-standing estimate of 140 Soviet long-range bombers has been revised upwards to 156 to conform with Soviet official data made public at the time of the signing of the SALT II Treaty. In past years, the designation 'Tu-20' has been given for the 'Bear' bomber in *SIPRI Yearbooks*. The SALT II Treaty states that the 'Bear' bomber is designated 'Tu-95' in the Soviet Union. Similarly, the Soviet designation for the medium-range bomber known in the West as 'Backfire' is referred to in the table as 'Tu-22M' (as opposed to 'Tu-26' in previous *SIPRI Yearbooks*) to conform with the designation used in the Soviet Backfire statement given to the USA before the signing of the SALT II Treaty.

US medium-range FB-111 strategic bombers are shown in parentheses, and long-range bombers only are included in the bomber totals, to clarify the number of delivery vehicles counted against SALT II limitations.

'Backfire' is included in the table only because much attention is given to this aircraft in the United States as a potential strategic delivery vehicle. It is the only weapon system in the table which is not officially recognized—indeed, disavowed—by the deploying government as a strategic weapon system. Moreover, it has been publicly recognized in US intelligence estimates as having less than intercontinental range in normal combat flight profile and as having been deployed at bases with peripherally oriented medium-range bombers and with naval aviation forces. As in the case of the Tu-95 'Bear', the naval aviation-assigned 'Backfires' are not included in the table at all. The medium-range bomber-assigned units, about half of production to date, shown in the table because of their prominence in the debate, are not included in the Soviet bomber totals.

For the past several years, the *Annual Report* of the US Secretary of Defense has included estimates of the total inventory of US bomber aircraft, including a large number of B-52s (about 220) in inactive storage. These aircraft will be counted against the SALT II delivery vehicle totals, even though many of them, perhaps most, are not in operating condition, and some may have been cannibalized or allowed to rust. (Almost all are older B-52 C/E/F models.)

Nuclear warheads

The estimates of independently targetable missile warheads can generally be reconciled with the official US estimates of total bomber and missile warheads if the following steps are taken: (*a*) bomber warhead loads are based on one bomb per 8 000–10 000 kg payload, using Unit

Equipment (UE) aircraft for the USA and adding SRAMs (1 140 operational missiles deployed on the bombers during 1972–75) to the internal payload; (*b*) in the case of US SLBMs, loads on submarines under conversion and in overhaul are excluded altogether; and (*c*) for some early years, individual MRVs and not just MIRVs are counted separately in the force loads total. The official US estimate of 7 000 independent nuclear warheads on Soviet strategic forces in 1981 can be obtained only if it is assumed that all Soviet MIRV-capable ICBMs are deployed with their maximum load and that some of the most recent 'Delta III' submarines have been deployed with a 7-warhead version of the SS-N-18 rather than the 3-warhead version shown in the table as deployed on 'Delta IIIs'.

5. Military use of outer space

Square-bracketed numbers, thus [1], *refer to the list of references on page* 113.

I. Introduction

The unique advantages of artificial Earth satellites circling the globe have now been exploited to a considerable degree, particularly for military purposes. The extensive use of spacecraft for various military missions is indicated in table 9.1 which shows the yearly launches of various spacecraft since 1958. This summary does not, however, indicate the many satellites launched for basic scientific measurements which may also be of considerable interest to the military. Some 75 per cent of all satellites are launched for military purposes; if we include a number of the scientific satellites, the proportion would be higher. The yearly launch rates of US and Soviet satellites, for most missions, have become constant during the last four years suggesting that the two powers now have as much satellite capacity as they want. Details of the military satellites launched during 1981 are given in the *SIPRI Yearbook 1982*, tables 9.2–9.9.

Advances made in military space technology include improved space-based sensors for surveillance, communications, command and control systems and space-based navigation aids to enhance the accuracies of delivery systems for both conventional and nuclear weapons. This advanced technology has contributed to refine war-fighting tactics. Over the past two decades or so nuclear war-fighting doctrines have evolved from mutual assured destruction (MAD) to the more unified concept of the countervailing strategy. While the former postulated a concept of massive retaliation against cities and industrial centres, the latter requires the maintenance of MAD capability as well as the ability "for flexible, controlled use of strategic weapons against appropriate targets for any attack at any level of conflict" [1]. The targets are mainly military rather than civilian. The new concept also assumes that "nuclear exchanges might not be quick exchanges but that they might last weeks or even months [2]".

High accuracies of weapon delivery systems, a precise knowledge of targets and adequate warning of attack are among the essential requirements of the new doctrine, while a flexible response capability requires secure communications, command, control and intelligence, so-called C^3I, systems. The extent to which satellites fulfil some of these requirements is briefly discussed below.

As the military has come to rely more and more upon satellites so their survivability has become increasingly threatened. Spacecraft are potential targets for anti-satellite (ASAT) systems. The current status of this aspect of space technology is briefly considered in section III.

II. The role of satellites in nuclear war strategy

The concepts of counterforce and countervailing strategies have only recently been publicly mentioned. But, at least in the USA, these ideas had been expressed as early as 1975 by the then Secretary of Defense James Schlesinger. He said that with "a reserve capability for threatening urban-industrial targets, with offensive systems capable of increased flexibility and discrimination in targeting, and with concomitant improvements in sensors, surveillance, and command-control, we could implement response options that cause far less civilian damage than would now be the case" [3].

Increased flexibility and discrimination in targeting are dependent upon an accurate knowledge of the targets and their locations. One of the factors which decrease collateral damage is the improved accuracy with which weapons could be delivered to their targets. Rapid transmission of targeting information and information directing the actions of the offensive forces needs command, control and communications systems. Space plays an essential role in these processes and the way in which the satellites perform their tasks is indicated below.

Reconnaissance satellites

If photographic, electronic and ocean surveillance and early-warning satellites are included in the reconnaissance satellite group, then they constitute about 50 per cent of all the military satellites launched during 1981 (see table 5.1).

Photographic reconnaissance satellites

The US budget for reconnaissance and surveillance from space is expected to be $1 180 million in FY 1982 and about $1 310 million in FY 1986 [4]. Considerable effort is being devoted to improving sensors such as infra-red devices and radars and to developing long-lived reconnaissance satellites. While both the USA and the USSR have such programmes, the former has actually deployed long-lived photographic reconnaissance satellites. For example, a US satellite (1978-60A) launched on 14 June 1978 had a lifetime of 1 166 days. While these satellites, known as KH-11 satellites, transmit images in real time in digital form to a ground station, the previous generation of spacecraft, the US Big Bird satellites, take photographs of the Earth's surface using high-resolution film cameras.

The films are returned to Earth for processing and analysis. The lifetimes of the Big Bird satellites have been about 180 days, but a more recent satellite (1980-52A) launched on 18 June 1980 decayed after 261 days.

Figure 5.1 shows the extent to which US and Soviet long-lived satellites have observed the earth since 1977. There is considerable overlap between the coverage of individual KH-11 satellites and that between KH-11 and Big Bird satellites. Such an overlap is also beginning to take place in the case of relatively long-lived Soviet satellites. It is interesting to ask why these overlaps occur, particularly when both types of satellite seem to be generating high resolution data. To some extent the answer can be deduced from the orbital characteristics of these satellites.

The conclusion is that, first, with more than one satellite in orbit the frequency of observation of any particular area is increased; whereas with one satellite the interval before the next observation is 92 minutes, with two satellites this time would be reduced considerably depending on the relative positions of the orbital planes. Second, because the orbits are spaced, a much larger part of the Earth's surface is covered at the same time. A satellite orbit can be fixed in space by two of the orbital elements,[1] the angle of inclination (i) of the orbital plane of the satellite to the earth's equatorial plane and the right ascension of the ascending node (Ω). (Detailed explanations of the various orbital elements can be found elsewhere [5, 6].) Here it is sufficient to note that in general all the orbital elements except i vary during the lifetime of a satellite. For an orbital inclination of 90°, however, the value of Ω does not change. For US reconnaissance satellites launched with an orbital inclination of about 97°, Ω does not change significantly.

The US satellites considered in figure 5.1 all have very similar orbital inclinations, that is, about 97°. The relative orientations of their orbital planes are, therefore, determined by Ω, the values of which are indicated in figure 5.1 for particular times.

The values of Ω for the KH-11 satellites 1976-125A and 1978-60A are seen from figure 5.1 to be very similar suggesting that satellite 1978-60A was probably a replacement for 1976-125A. The overlap between KH-11s 1978-60A and 1980-10A is for a very long period of time. The difference in Ω for these satellites is 46°. A similar difference (49°) in the values of Ω can be observed between KH-11s 1980-10A and 1981-85A. The latter satellite was launched some 10 days after 1978-60A had decayed. The difference in the values of Ω for these two satellites is not very large (about 9°) suggesting that 1981-85A may have been a replacement for 1978-60A. Since the launch of satellite

[1] The orbital elements are a set of six parameters defining the orbit of a satellite. These are the right ascension of the ascending node (Ω), the orbital inclination (i), the argument of the perigee (ω), the semi-major axis of the orbit (a), the eccentricity of the orbit (e) and the time of perigee passage (T).

Figure 5.1. Coverage by US and Soviet long-lived photographic reconnaissance satellites launched during 1977–81

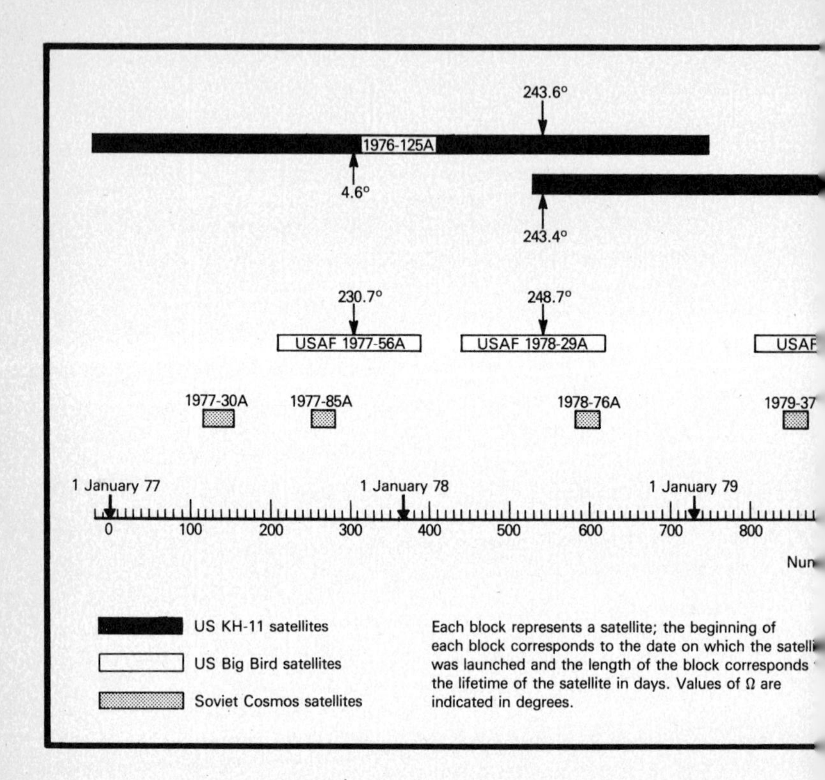

1980-10A in early 1980, with the exception of some 10 days in 1981, two KH-11 reconnaissance satellites have been in orbit at the same time which suggests that a new two-satellite pattern has been established. This is apparent from figure 9.2 in which the ground tracks obtained over a period of 24 hours for satellites 1980-10A and 1981-85A are plotted. From the figure it can be seen that the gap between two consecutive tracks of satellite 1980-10A is filled in by a ground track of satellite 1981-85A thus increasing the frequency of observation. A wider coverage of the Earth at any given time is also possible because the orbits are spaced (in this case by 49° and in the case of KH-11s 1978-60A and 1980-10A by some 46°) such that a pair of satellites can observe different parts of the Earth's surface simultaneously.

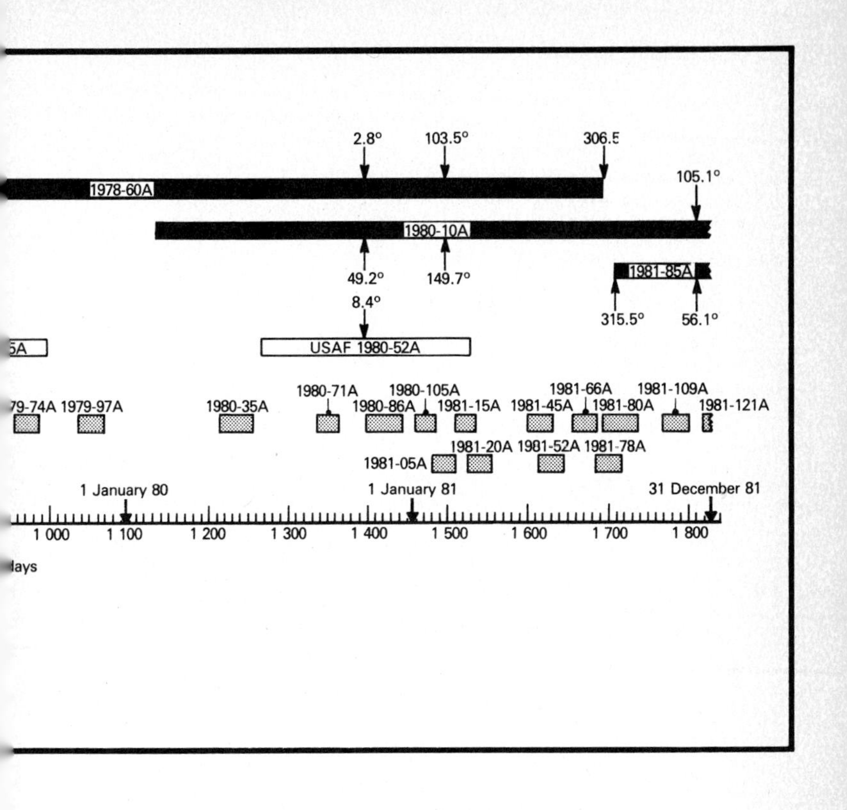

From the values of Ω (see figure 9.1) no clear pattern emerges in the relationship between KH-11 and Big Bird satellites. This may be because KH-11 satellites are operated by the CIA while the US Air Force is in charge of the Big Bird satellites. It is, however, possible that the Big Bird satellites are used to obtain photographs of areas which the KH-11 spacecraft indicate to be of specific interest.

While the sensors on board the KH-11 satellites are sophisticated, highest quality images can probably still be obtained only by using photographic equipment. The Big Bird satellites carry such equipment. In fact, there have been reports suggesting that photographs taken from US satellites have shown a new Soviet variable-geometry swept-wing aircraft [7]. Such are the details observable from space.

The above analysis is confined to US satellites: it is not possible to make similar observations of Soviet satellites because the USSR has still not launched any very long-lived satellites. The majority of Soviet satellites have a lifetime of 14 days. In 1981, however, 10 long-lived satellites were launched, most having lifetimes of about 30 days and three with lifetimes of over 40 days. This was five more long-lived satellites than in 1980. Moreover, since the Soviet reconnaissance satellites are launched at an orbital inclination of about 67°, the rate of change of Ω is not zero. In any case, the extent of overlap is relatively small. The pattern in which these satellites have been launched can be seen from figure 5.1.

Electronic reconnaissance satellites

While much is known about photographic reconnaissance satellites, knowledge of electronic surveillance spacecraft is comparatively scant. Clearly, such satellites act as ears in space for the military. They are designed to detect and monitor radio signals generated by the enemy's military activities both within their country and throughout the world. Signals originate from military communications between bases, from early-warning radars, air-defence and missile-defence radars or from those used for missile control. These satellites also gather data on missile testing, new radars and many other types of communications traffic.

It is important to locate precisely the sources of the signals intercepted by electronic reconnaissance satellites. This task could be performed by navigation satellites [8] by a method resembling hyperbolic navigation [9]. The process is reversed and instead of using four transmitters on four satellites to establish the position of a known object, four receivers on four satellites are used to locate the position of an unknown transmitter on the Earth's surface.

Both the USA and the USSR are known to have launched electronic reconnaissance satellites. The US Air Force introduced the first generation of such satellites in 1962 while the first Soviet electronic reconnaissance satellite was launched in 1967. A difference is that, while the US electronic satellites are mainly launched on board Big Bird satellites, the Soviet satellites are still launched by means of independent launchers. The Big Bird passengers are ejected into independent orbits at much greater altitudes. No such satellite was launched in 1981.

In recent years the Soviet Union has launched electronic reconnaissance satellites at orbital inclinations of 74° and with orbital periods of 95.2 minutes. Those launched at orbital inclinations of 81° and with periods of about 97.6 minutes were previously thought to be meteorological satellites, but these are now thought to belong to the electronic reconnaissance series of satellites. A satellite launched in 1981, Cosmos 1311,

Figure 5.2. **Southbound ground tracks of two KH-11 satellites (1980-10A and 1981-85A) for a period of 24 hours, showing the frequency of observation when two satellites are used**

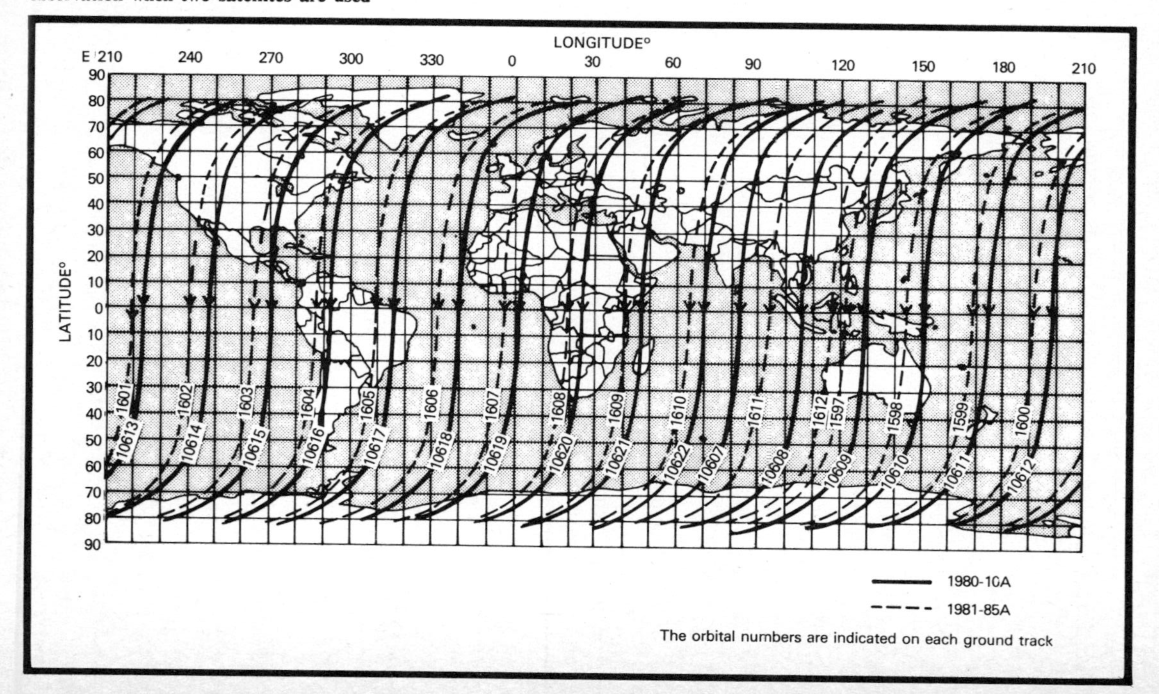

at an orbital inclination of about 83° with a period of 94.5 minutes may be the start of a new series [10].

Besides these electronic reconnaissance satellites, the Soviet Union appears to be using others with orbital inclinations of 65° and orbital periods of 93.3 minutes for ocean surveillance. These satellites detect and monitor radio transmissions and radar signatures probably originating from naval surface ships [10].

Other satellites

Other satellites in the intelligence part of the C^3I system are ocean surveillance and oceanographic satellites, early-warning satellites and nuclear explosion detection satellites. Ocean surveillance satellites detect and track military surface ships while oceanographic satellites are used to determine various ocean properties in order to enhance understanding of the behaviour of sound in oceans, a knowledge of which would increase anti-submarine warfare capabilities. The early-warning satellites detect missiles soon after they are launched.

The USA has launched satellites for the specific detection of nuclear explosions in the atmosphere and in outer space. However, it is now planned that NAVSTAR will carry sensors for this mission under the Integrated Operational Nuclear Detection System (IONDS). This system is intended to provide damage assessment both within one's own country and within enemy territories during and after a nuclear attack. This effort is to support the new nuclear war doctrine which requires early warning of attack, information for assessing the size of the attack and data on the attacked target so that an appropriate response could be made [2].

Communications satellites

Communications satellites are only part of an elaborate communications system. However, in the USA the Department of Defense has preferred satellite communications and has proposed improved communications with the nuclear forces [2]. The existing space component in the US communications system consists of communications transponders aboard satellites such as the Fleet Satellite Communications (FLTSATCOM) satellites, as well as other satellites in polar orbits. The latter satellites are, for example, the Defense Satellite Communications System and the Satellite Data System satellites.

Since satellites are becoming an essential element of the C^3I system, it has become important to make them survivable. The US Air Force has proposed a satellite for this specific purpose, called STRATSAT, which would orbit at an altitude of about 203 720 km in order to increase its survivability. To improve its resistance to electronic jamming and disturbance of the electromagnetic environment, the satellite would use extra high frequencies and sophisticated electronic methods. It is also proposed that the satellite should have manoeuvring capabilities [2]. While these measures may make satellites survivable and operable, the ground segment of the system, and to some extent the satellites themselves, may suffer considerable damage from the effects of nuclear detonations. Electronic systems could be disrupted without a direct hit by a nuclear warhead.

A nuclear detonation produces, among other effects, a pulse of high-energy gamma rays which would mainly affect semi-conductor devices, a pulse of neutrons which would significantly and permanently alter the electrical properties of semi-conductor devices and an electromagnetic pulse (EMP) which would produce very high voltages and currents in cables, metal enclosures and structures thus causing breakdown in insulations and the destruction of electronic circuits. The electromagnetic pulse would affect very wide areas on the Earth's surface if a nuclear weapon were detonated above the atmosphere, and electronic systems would be incapacitated.

This damage could be reduced to some extent by hardening electronic components at the design stage. Considerable efforts are being devoted to harden both the ground and the space segment of the C^3I [2].

III. ASAT systems

While on one hand considerable resources are devoted to making one's own satellites survivable, much is being done to develop methods of destroying the enemy's spacecraft. In fact the Soviet Union launched three test satellites in 1981; one was a target satellite and the other two were interceptors. These tests were of the co-orbital type in which the interceptors were almost in the same orbital plane as the target. The interceptors approached the target slowly, each coming within a distance of 10 km before de-orbiting and re-entering the earth's atmosphere. In other tests in the past the Soviet Union has orbited the target and the interceptor in different orbital planes, with the interceptor in an eccentric rather than a circular orbit. In this case the interceptor was exploded after the interception. In a third method, the interceptor ascends close to the target and the interception is made before the interceptor has completed a full orbit. The interceptor is then commanded back to Earth.

The USA is planning to begin operational testing of its ASAT system in 1983 and may achieve initial operational status by 1985 [11]. The system consists of a two-stage solid-propellant missile carrying a non-nuclear warhead called the miniature homing vehicle (MHV) designed to damage the enemy satellite. Such missiles will be launched from F-15 aircraft flying at an altitude of some 20 km.

Initial testing of this system on the ground has already begun [12]. The MHV is just under 5.5 m long, 0.5 m in diameter and weighs about 1 180 kg. The first stage of the missile is a booster of a short-range attack missile and the second stage is a smaller Altair 3 solid rocket motor. The MHV is mounted on the frame of the second stage of the missile. Since it is not confined to any specific launch pad on earth this ASAT system has considerable advantages over the Soviet system. It can be used as a direct ascent system against satellites in any orbital inclination. This flexibility together with the small size of the MHV increases its survivability.

The MHV will be guided to its target by an infra-red homing device. It has recently been reported that the Soviet Union has deployed such infra-red homing interceptors on an anti-satellite battle station in low earth orbit [13], but the Pentagon has denied this possibility [14].

Besides these systems, both the USA and the USSR are investigating high-energy laser and particle beams for ASAT applications. For example, in 1982 the USA is expected to spend $279 million on research with the intention of producing high-energy laser weapons [15]. The Department of Defense has been working on laser weapons for over a decade and by the end of fiscal year 1981 will have spent about $1 500 million on investigations into laser weapons. Similar efforts must have been carried out in the USSR.

In early February 1981, the US Air Force tested its aircraft-mounted laser weapon at full power on the ground [16]. Later, on 1 June, the airborne laser weapon was tested against an air-launched AIM-9L Sidewinder air-to-air missile. While the beam hit the target, it did not successfully destroy it [17]. This Airborne Laser Laboratory is equipped with a 400-kW gas dynamic laser operating at a wavelength of 10.6 μm in the infra-red region of the electromagnetic spectrum. Two days later a second test was carried out against an AIM-9L missile. It was reported that this second test had achieved better results because the beam was able to lock on to the target for a long period [18]. While there are a number of technical problems to be solved it appears that, at least in the USA, the nature of laser beam propagation is reasonably well understood [15]. The US Defense Advanced Research Projects Agency is now going to focus its study on the effects of high-energy laser beams on targets such as aircraft and other types of vehicle. The Department of Defense is studying the high-energy laser weapons under two demonstration projects: one is the US Air Force's Airborne Laser Laboratory and the other is the Navy's Sea Lite. The latter uses a chemical laser considerably more powerful than the device used by the former.

IV. Control of the militarization of outer space

From this brief description of the development of military activities in outer space, it is not surprising that both the USSR and the USA have shown at least some interest in limiting their ASAT activities. Both parties, in fact, met once in 1978, then in January 1979 and again from April to June 1979 to discuss the control of their ASAT programmes. The discussions did not seem to have been very productive and no further meetings have taken place.

However, since these talks, a significant move made by the Soviet Union in 1981 appears to have removed the discussions from the bilateral forum to the multilateral one. In August 1981, the USSR proposed to the United Nations a new treaty banning the placement of any kind of weapon into orbit around the Earth [19].

However, while the proposal clearly bans the deployment of ASAT weapons in orbit, it does not ban such weapons within the atmosphere and above it. Examples of such weapons are the US MHV system and the Soviet ASAT satellites which do not complete an orbit. While these are some of the drawbacks of the Soviet proposal, nevertheless, it contains some far-reaching measures compared to the 1967 Outer Space Treaty.

In its verification clause the proposal states that "each state party shall use the national technical means of verification at its disposal" to provide assurances of compliance with the provisions of the treaty. Since only two nations today possess the necessary technological base needed for verification, however, it is difficult to visualize many nations becoming parties to the treaty unless an international verification agency is created, as was proposed by France in 1978.

During the 1978 special session of the United Nations General Assembly on disarmament, France proposed the setting up of an International Satellite Monitoring Agency (ISMA). The General Assembly requested the Secretary-General to undertake, with the assistance of qualified governmental experts, a study on the technical, legal and financial implications of establishing an ISMA. The results of the study have since been published [20]. The main conclusions of the report are that: (*a*) space technology would allow observations from satellites for the verification of compliance with arms control and disarmament treaties and for monitoring crisis areas on Earth; (*b*) there is no provision in any international law that would prevent an international government agency from carrying out observations by satellite; and (*c*) the financial burden of the agency in its final phase, when it launches and operates its own satellites and carries out data processing and analysis, is expected to be about $1 500 million (for one satellite) spread over a 10-year period. In any case the annual cost of an ISMA to the international community would be very much less than 1 per cent of the total yearly expenditure on armaments.

The first conclusion is based on the fact that the capabilities of civilian space technology for observing the Earth's surface are beginning to approach those of military technology in many respects. Moreover, satellite technology is spreading into many more countries and launcher technology has reached countries not otherwise considered to be very advanced in this field (see table 5.2). More importantly these countries are also acquiring the technology for image processing, essential for the interpretation of data from space [21]. Undoubtedly these trends will continue and space technology will spread beyond the industrial nations. Once many states are able to observe the Earth from space, the fear of releasing sensitive data—one of the most serious objections to establishing an ISMA—may no longer be relevant.

There are a number of issues to be resolved before an ISMA could be created. Verification could not be carried out from space alone and data from other sources would be necessary. A number of existing international organizations could be involved in the verification of some specific arms control/disarmament treaty, such as the World Health Organization, the World Meteorological Organization, the International Atomic Energy Agency and the International Telecommunications Union [21]. Difficult questions concerning the modalities of data acquisition and dissemination, of direct relevance to the sensitive security considerations of states, must be dealt with. However, solutions to such problems will not be found unless discussions continue. These discussions may even consider the possibilities of a verification agency on a multinational or regional basis. For example, once the Ariane launcher becomes routinely available, the European Space Agency (ESA)[2] could contribute to the verification of any arms control measures that may be worked out in Europe. In this context Interkosmos[3] could also have a vital role to play.

V. Discussions

Both the Soviet and the US military authorities are beginning to depend heavily on artificial Earth satellites. Successful launches of the US reusable

[2] Belgium, Denmark, France, the Federal Republic of Germany, Ireland, Italy, the Netherlands, Norway, Spain, Sweden, Switzerland and the UK are members of the European Space Agency.

[3] Bulgaria, Cuba, Czechoslovakia, the German Democratic Republic, Hungary, Mongolia, Poland, Romania, and the USSR are members of Interkosmos.

launcher, the Space Transportation System or space shuttle, in 1981 have opened up the door for further proliferation of military activities in outer space. The US Air Force has suggested it should have more control over the future activities of the space shuttle [22]. It has also been suggested that under the Block II shuttle programme the capabilities of the space shuttle should be increased by enlarging the vehicle to accommodate larger payload.

The second shuttle launch, *Columbia 2*, on 12 November 1981 was a significant one. First, it showed that a space transportation vehicle could be used again and, second, it carried a number of test payloads which increased the mass lifted into orbit relative to the previous launch in April 1981. The test payloads included a manipulator arm to be used later for placing satellites into orbit and retrieving satellites from orbit for inspections and repair. Under the OSTA-1 programme (Office of Space and Terrestial Applications) the Shuttle Image Radar-A (SIR-A) and Shuttle Multispectral Infra-red Radiometer (SMIRA) were also carried. *Columbia 2* was placed in a circular orbit at an altitude of about 250 km with an orbital inclination of 38°. It orbited in an inverted position with its cargo bay doors facing the Earth and open. The SIR-A and the SMIRA functioned successfully [23]. The ground resolution of SIR-A, the first side-looking radar to be orbited, is expected to be about 80 m (or an instantaneous field of view (IFOV) of 40 m × 40 m) [24].

During 1981, the virtual monopoly of the USA and the USSR for launching spacecraft was broken by the European Space Agency's successful launch (using a single Ariane launcher) of two satellites, the European Weather Satellite, *Meteosat*, and the Indian communications satellite, *Apple*. In 1981, the People's Republic of China also launched three satellites using a single launcher. This came after a lull of some three years.

Improvements in space-based sensors for surveillance, communications, command and control systems and space-based navigation technology to enhance the accuracies of delivery vehicles for both conventional and nuclear weapons are relevant not only to the current nuclear arms race but also to war-fighting strategies. This link between the arms race on earth and space technology is further emphasized by the fact that nuclear weapon states orbit spacecraft by means of launchers based on missiles developed to carry nuclear warheads. In a separate trend some countries outside the group of nuclear weapon states have developed launchers primarily for orbiting satellites. It will be an opportunity missed if greater control over the militarization of space is not brought about now. In this context it is essential that discussions of concepts such as ISMA be kept alive and an improved outer space treaty be worked out.

VI. Tables

Table 5.1. Summary of possible military satellites by type of mission

	Photographic reconnaissance satellites			Electronic reconnaissance satellites		US MIDAS and Vela satellites		Early-warning satellites		Ocean-surveillance satellites		Navigation satellites	
Year	USA	USSR	China	USA	USSR	MIDAS	Vela	USA	USSR	USA	USSR	USA	USSR
1958	–	–	–	–	–	–	–	–	–	–	–	–	–
1959	6	–	–	–	–	–	–	–	–	–	–	1	–
1960	6	–	–	–	–	2	–	–	–	–	–	2	–
1961	13	–	–	–	–	3	–	–	–	–	–	3	–
1962	26	5	–	4	–	1	–	–	–	–	–	1	–
1963	17	7	–	7	–	2	2	–	–	–	–	3	–
1964	24	12	–	8	–	–	2	–	–	–	–	3	–
1965	21	17	–	5	–	–	2	–	–	–	–	4	–
1966	23	21	–	10	–	2	–	1	–	–	–	4	–
1967	18	22	–	8	5	–	2	–	1	–	1	3	–
1968	16	29	–	7	7	–	–	1	1	–	1	1	–
1969	12	32	–	6	11	–	2	1	–	–	–	–	–
1970	9	29	–	7	10	–	2	3	–	–	1	1	1
1971	7	28	–	3	15	–	–	1	–	4	2	–	2
1972	8	30	–	3	7	–	–	2	1	–	1	1	3
1973	5	35	–	2	12	–	–	2	1	–	1	1	3
1974	5	28	–	3	10	–	–	–	1	–	2	1	4
1975	4	34	1	2	8	–	–	2	2	1	3	1	4
1976	4	34	1	1	11	–	–	1	1	4	2	1	8
1977	3	33	–	–	8	–	–	2	3	4	3	1	8
1978	2	35	1	1	6	–	–	2	2	1	–	4	8
1979	2	35	–	1	5	–	–	2	2	–	3	–	6
1980	2	35	–	1	6	–	–	–	5	4	4	2	6
1981	2	37	–	–	4	–	–	2	5	–	8	1	5
Total by country	**235**	**538**	**3**	**79**	**125**	**10**	**12**	**22**	**25**	**18**	**32**	**39**	**58**
Total by mission		**776**		**204**		**22**		**47**		**50**		**97**	

[a] Fractional orbital bombardment system.

Communications satellites					Meteorological satellites				Geodetic satellites			FOBSs[a]	Interceptor/ destructor satellites		
USA	USSR	NATO	UK	France	USA	USSR	France	UK	USA	USSR	France	USSR	USSR	Yearly total	Cumulative total
1	–	–	–	–	–	–	–	–	–	–	–	–	–	1	1
–	–	–	–	–	–	–	–	–	–	–	–	–	–	7	8
2	–	–	–	–	2	–	–	–	–	–	–	–	–	14	22
2	–	–	–	–	1	–	–	–	–	–	–	–	–	22	44
3	–	–	–	–	4	–	–	–	1	–	–	–	–	45	89
4	–	–	–	–	3	2	–	–	–	–	–	–	–	47	136
3	3	–	–	–	3	2	–	–	2	–	–	–	–	62	198
7	8	–	–	–	6	4	–	–	6	–	–	–	–	80	278
11	2	–	–	–	6	2	–	–	4	–	1	2	–	89	367
17	5	–	–	–	6	4	–	–	1	–	2	9	1	105	472
11	4	–	–	–	4	2	–	–	1	2	–	2	4	93	565
5	2	–	1	–	3	2	–	–	1	2	–	1	2	83	648
3	14	1	1	–	5	6	–	–	1	–	1	2	3	100	748
5	21	1	–	–	2	4	1	1	–	2	–	1	6	106	854
3	24	–	–	–	4	5	–	–	–	2	–	–	–	94	948
4	33	–	–	–	2	3	–	–	–	1	–	–	–	105	1 053
3	24	–	2	1	4	6	–	–	–	2	–	–	–	96	1 149
5	37	–	–	1	3	5	–	–	1	2	1	–	–	117	1 266
11	29	1	–	–	3	3	–	–	1	1	–	–	1	118	1 384
4	16	1	–	–	2	3	–	–	–	1	–	–	7	99	1 483
6	42	1	–	–	4	–	–	–	–	1	–	–	1	117	1 600
3	27	–	–	–	2	4	–	–	–	–	–	–	2	94	1 694
3	36	–	–	–	2	2	–	–	–	–	–	–	3	111	1 805
2	39	–	–	–	2	2	–	–	–	–	–	–	3	112	1 917
118	**366**	**5**	**4**	**2**	**73**	**61**	**1**	**1**	**19**	**16**	**5**	**17**	**33**	**1 917**	
		495				**134**				**40**		**17**	**33**	**1 917**	

Table 5.2. Present status of spacecraft launchers

Country	Launcher name	No. of stages	Type of fuel	Payload (kg) Low orbit (200–1 000 km)	Payload (kg) Synchronous orbit	Cost of launching (million dollars)	First flight
France	Diamant BP-4	4	Liquid and solid	140	..	..	1970
ESA	Ariane	3	Liquid	5 900	526	16	1979
India	SLV-3	4	Solid	40	..	..	1978
Japan	L-4S	4	Solid	12	..	..	1966
	M-4S	4	Solid	75	..	..	1971
	M-3C	3	Solid	160	..	..	1974
	M-3H/ M-3S	3	Solid	270	..	..	..
	N-1	3	Liquid and solid	400	..	..	1975
	N-2	3	Liquid and solid	1 100	..	..	1981
USA	Scout	4	Solid	200	..	5.2	1960
	Delta	3	Liquid and solid	2 040	400	9.2	1960
	Atlas/ Centaur	2	Liquid	4 900	1 800	18.7	1962
	Titan 3C	4	Liquid and solid	11 340	1 450	23.2	1965
	Titan 3D	2	Liquid and solid	13 600	..	..	1971
	Titan 34D	3	Liquid and solid	14 900	1 900	..	Planned for 1981
	Titan 3E/ Centaur	4	Liquid and solid	13 600	3 530	29.3	1974
	Shuttle	2	Solid	1 590	2 270	12–15	1979
USSR	A (SS-6 Sapwood)	2	Liquid	..	..	..	1957
	B (SS-4 Sandal)	2	Liquid	300–420	..	..	1962
	C (SS-5 Skean)	2	..	500–1 000	..	..	1964
	D (Proton)	4	..	13 000–22 500	..	..	1965
	F (SS-9 Scarp)	3	..	2 500–4 700	..	..	1966

References

1. *Countervailing Strategy Demands Revision of Strategic Force Acquisition Plans*, Report No. MASAD-81-35 (U.S. General Accounting Office, Washington, D.C., 5 August 1981).
2. *Strategic Command, Control and Communications: Alternative Approaches for Modernization, A CBO Study* (Congressional Budget Office, Washington, D.C., October 1981).
3. *Annual Defense Department Report, FY 1975*, Report of the Secretary of Defense James R. Schlesinger to the Congress on the FY 1975 Defense Budget and FY 1975–1979 Defense Program (U.S. Department of Defense, Washington, D.C., 4 March 1975).
4. 'Military reconnaissance & surveillance market projected at $19.4 B for FY 1986', *Journal of Electronic Defense*, Vol. 4, No. 4, July/August 1981, p. 9.
5. SIPRI, *Outer Space—Battlefield of the Future?* (Taylor & Francis, London, 1978), pp. 4–20.
6. SIPRI, *World Armaments and Disarmament, SIPRI Yearbook 1975* (Almqvist & Wiksell, Stockholm, 1975), pp. 378–91.
7. 'New Soviet bomber photographed by U.S. satellite', *Defense Daily*, Vol. 119, No. 30, December 1981, p. 238.
8. Melton, W. C., 'Time of arrival measurement possible with use of Global Positioning System', *Defense Electronics*, Vol. 13, No. 12, December 1981, pp. 90–98.
9. SIPRI, *World Armaments and Disarmament, SIPRI Yearbook 1981* (Taylor & Francis, London, 1981), pp. 127–30.
10. Perry, G. E., 'Identification of military components within the Soviet space programme', in *Outer Space—A New Dimension of the Arms Race*, ed. B. Jasani (Taylor & Francis, London 1982, Stockholm International Peace Research Institute), to be published.
11. Ulsamer, E., 'Go-ahead on USAF's ASAT Program', *Air Force Magazine*, Vol. 64, No. 10, October 1981, p. 16.
12. Smith, B. A., 'Vought tests small antisatellite system', *Aviation Week & Space Technology*, Vol. 115, No. 19, 9 November 1981, pp. 24–25.
13. 'Killer satellites', *Aviation Week & Space Technology*, Vol. 115, No. 17, 26 October 1981, p. 15.
14. 'Pentagon denies Soviet orbital battle station', *Flight International*, Vol. 120, No. 3783, 7 November 1981, p. 1363.
15. 'DARPA Chief assesses laser weapon program', *Laser Focus*, Vol. 17, No. 12, December 1981, pp. 34–38.
16. 'USAF tests high-energy laser weapon', *Flight International*, Vol. 119, No. 3744, 7 February 1981, p. 334.
17. 'Laser fails to destroy missile', *Aviation Week & Space Technology*, Vol. 114, No. 23, 8 June 1981, p. 63.
18. 'Second laser laboratory test', *Interavia Air Letter*, No. 9778, 26 June 1981, p. 8.
19. UN document A/36/192, 20 August 1981.
20. UN document A/AC.206/14, 6 August 1981.
21. Jasani, B. and Karkoszka, A., 'International verification of arms control agreements', paper presented to the Independent Commission on Disarmament and Security Issues under the Chairmanship of Olof Palme, December 1981.

22. 'USAF needs space command for shuttle', *Flight International*, Vol. 120, No. 3787, 5 December 1981, p. 1674.
23. 'Second shuttle mission was a success', *Interavia Air Letter*, No. 9896, 10 December 1981, pp. 6–7.
24. Taranik, J. V. and Settle, M., 'Space shuttle: a new era in terrestrial remote sensing', *Science*, Vol. 214, No. 4521, 6 November 1981, pp. 619–26.

6. The neutron bomb

Square-bracketed numbers, thus [1], *refer to the list of references on page* 126.

I. What is a neutron bomb?

The 'neutron bomb' is the popular name given to those nuclear weapons whose predominant effect is to cause casualties from the neutrons emitted when it is exploded. Any very low-yield weapon (less than about 1 kiloton), whether its energy is derived from fission or fusion reactions, has this property. The neutron effects can, however, be enhanced if fusion reactions are used to produce the energy, since these produce large numbers of high-energy neutrons, which are not so easily absorbed by bomb materials and air. They can also be enhanced by constructing the weapon so that fewer neutrons are trapped in its outer layers. Hence in official circles 'neutron bombs' are normally referred to as enhanced radiation weapons (ERWs) because the weapons are specifically designed to produce more neutrons per unit of energy released and to allow these neutrons to escape from the surrounding bomb materials.

No matter how they are designed, high-yield weapons will never be 'neutron bombs'; conversely, for all very low-yield weapons the blast effects will be relatively unimportant compared to the neutrons. As the yield is reduced, first thermal radiation and then blast become less important in comparison with prompt gamma radiation and then neutrons. This is because the scaling laws for calculating how the range of a given effect varies with the yield are different for thermal, blast, gamma radiation and neutrons. Neutrons and gamma radiation are strongly absorbed by the air at normal atmospheric pressures so that as a consequence they do not travel long distances at low altitudes. Since neutrons are slightly more strongly absorbed than gamma rays, the range of neutron effects increases with yield even more slowly than that of gamma rays. Blast waves are only slightly attenuated as they propagate through the atmosphere, and the range for blast effects increases as the cube root of the yield. Thus in the case of higher-yield weapons they extend far beyond those from neutrons and gamma rays. Thermal radiation is not absorbed significantly in clear air and, therefore, the range increases as the square root of the yield. Thus for very high-yield explosions (megatons) in relatively clear weather, thermal radiation can predominate over blast, and prompt nuclear radiations (neutrons and gamma rays) are inconsequential; for very low-yield weapons the reverse situation exists. The cross-over yield below which

Figure 6.1. Scaling laws for weapon effects

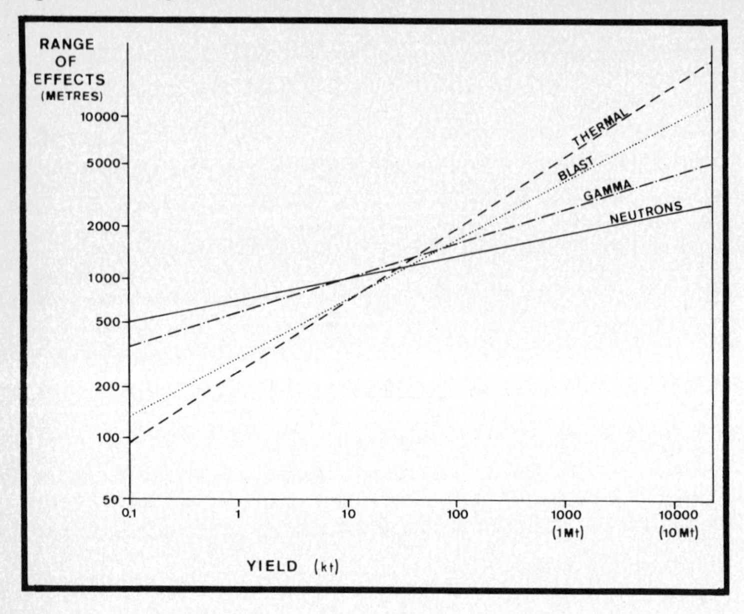

neutrons become predominant is about 1 to 10 kt. A representative diagram showing these relationships is given in figure 6.1.

By the use of enhancement techniques it is possible to design a 1-kt weapon that will have nuclear radiation effects on tank crews at about the same distance as a 10-kt standard fission weapon. Since the total yield will, however, be only one-tenth as great, the distance at which blast effects will be produced will be decreased by the cube root of 10 or slightly more than a factor of two. This neutron enhancement is obtained by having only a very low-yield fission trigger and obtaining most of the energy from fusion reactions. Furthermore, all materials which are good absorbers of high-energy neutrons are eliminated from the outer layers of weapons. In particular natural uranium, which would normally be included in a fission weapon to capture any high-energy neutrons and produce additional fission and energy, is eliminated. Thus more high-energy neutrons can escape into the atmosphere.

The partition of energy from a standard fission and an enhanced radiation warhead is shown in figure 6.2. Because the neutrons can escape more easily from the ERW materials, the fraction of energy in prompt radiation is much higher (approximately 30 per cent as compared with 5 per cent).

Figure 6.2. Typical energy partition

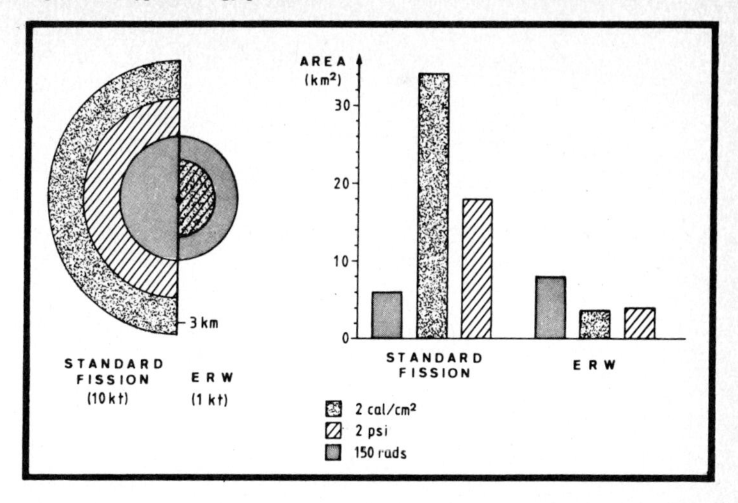

Since these neutrons are not absorbed directly around the bomb there is less heating and the amounts of energy that come out as blast and thermal radiation are each about 10 per cent lower. It will be noted, however, that 40 per cent of the energy is still emitted as blast so that fission weapons and ERWs with the same yield would not have very significant differences in the radii for blast damage. The lower yield of the enhanced radiation warhead, rather than its enhancement characteristics, accounts for the reduced blast damage.

II. What are the security justifications for the neutron bomb?

The ostensible military justification for the neutron bomb is to improve NATO capabilities to repel Warsaw Treaty Organization (WTO) massed tank attacks. The WTO outnumbers NATO in numbers of tanks, and a persistent fear of Western military planners has been that Soviet tank columns could break through the West German front and wreak havoc on NATO defences. NATO has very formidable anti-tank weapons, including high technology precision-guided munitions (PGMs) using non-nuclear kill mechanisms, but military planners have argued that this is not enough. As a consequence nuclear artillery shells and short-range missiles have been deployed for many years to force the USSR to disperse its tank forces. The enhanced radiation warheads are modernized versions of these, which it is hoped would render Soviet tank superiority ineffective.

One concern about existing Western nuclear warheads has been the collateral damage they would cause if they were actually used in the built-up areas of Western Europe. It was recognized that in the process of repelling Soviet tanks, NATO might end up by destroying Western Europe. These collateral effects might prevent Western leaders from ever authorizing their use. If the Soviet leaders could rely on this, then they might not be deterred from aggression with massed tank formations. The ERWs, because of the reduced blast effects of their low yields, were sought as a means to increase the credibility of the deterrent against such Soviet aggression. Making the decision to use nuclear weapons easier would thus make it less likely that they would actually be used.

III. Current status of the neutron bomb programme

In the 1960s the US Army had deployed in Europe a small bazooka-like nuclear weapon which could be handled by two men and had a yield of a few hundredths of a kiloton. Although this was a fission device with no enhancement characteristics, its primary effects would have been to kill personnel with neutrons. This was deployed without any public attention and then later withdrawn because the Army did not find it militarily useful.

In the late 1970s the United States developed enhanced radiation warheads for the Lance missile with a range of about 100 km, and for the 203-mm (8-inch) artillery howitzer with a range of 29 km. This was part of the modernization programme for NATO weapons. Consideration was also given to the development of such a warhead for the short-range, 155-mm howitzer, but this apparently ran into some technical difficulties because of the small diameter. While these are reported to have been solved, the warhead is not believed to be in the approved production programme. In 1978 President Carter approved production of the non-nuclear but not the nuclear components for the Lance and 203-mm shells. In the summer of 1981 President Reagan authorized, without consultation with his NATO allies, the procurement and stockpiling of the complete enhanced radiation warheads, but in order to mollify public opposition in Europe he announced they would not be deployed overseas at this time. If they are to be a deterrent to a Soviet tank attack, they will, however, have to be moved to Europe well in advance of a crisis. Thus the decision to deploy them in Europe cannot be avoided forever.

The old fission warhead for the Lance missile is reported to have the option of having several yields—1 kt, 10 kt and 70 kt. The 10-kt version is believed to be the optimum yield for forcing dispersal of Soviet tanks. The new enhanced radiation warhead would have a yield of about 1 kt.

The 203-mm howitzer fission warhead reportedly had yield options of 1–2 kt, and the new enhanced radiation version presumably a somewhat lower yield, reportedly 0.7 kt.

IV. The effects of nuclear weapons [1]

Nuclear radiation

The nuclear radiations—neutrons and gamma rays—do not destroy tanks, but only incapacitate the crews manning them. Their effects on people are, however, complex, and the onset of symptoms varies widely with the total dose or exposure. Neutron effects are different from gamma rays, but because there is less experimental data, they are not nearly as well known. In an actual conflict people will often be exposed to a mixture of gamma rays and both high- and low-energy neutrons, whose biological effects are different. Attempts are made to take these complications into account in military analyses, but it must be realized that there will never be a sharp line dividing the exposure which will incapacitate a soldier or injure a civilian bystander.

A person exposed to a lethal dose of approximately 450 rads, for example, would not be immediately put out of action but could become sick several hours to a day later and would die within a month. If the exposure is increased to about 2 000 rads, then the onset of symptoms will be much earlier. A person might be temporarily incapacitated almost immediately, suffering shock and perhaps even nausea, but could then temporarily recover for a period of several hours, then relapse and die a few days later. If the exposures were even higher, say 8 000 rads or over, then an individual would be put out of action almost immediately and die in a relatively short period of time [2].

The normal metals in a tank are not particularly effective in absorbing neutrons so they do not provide any significant protection to the crew. Special neutron shielding in a tank may or may not be practical. The steel in a tank would, however, reduce the exposure to gamma radiation, which becomes relatively more important compared to neutrons as the yield of the warhead increases. Taking all these factors into account the US Defense Department has used for analysis purposes an exposure of 3 000 rads as sufficient to put a tank out of action.

Much lower exposures will have significant longer-term effects. These will be of no immediate military use but must be taken into consideration in evaluating the consequences of the use of these weapons on military personnel and civilian bystanders. An exposure of 100–200 rads could produce early symptoms, particularly nausea and lowered blood counts. Exposure to 25 rads could significantly increase the risk of leukaemia several years later. Since large populations will be exposed in a war, even lower doses might cause significant numbers of people to suffer some long-term effects.

Blast effects

To prevent the tank itself from functioning, the US Defense Department has estimated that it would require a blast pressure of approximately 17 psi (lb/in^2). Under this type of exposure the tank would no longer be operable even with a healthy crew. Its treads could be damaged or the tank even be rolled over. The damage should be visible to an observer from some distance away.

Much lower pressures are required to damage buildings. For example, approximately 3 psi would be sufficient to produce significant damage to civilian structures, and there would be considerable destruction at even lower pressures. A blast pressure of 6 psi would seriously damage most civilian structures and the 17 psi needed to knock out tanks would leave few buildings standing.

V. Effectiveness of nuclear weapons against tanks

To evaluate the usefulness of the neutron bomb, the actual effects must be examined in detail. For this analysis a 1-kt ERW was selected as a typical neutron bomb, although the actual yields of the weapons deployed might range from about 0.5 to 5 kt. For comparison a 10-kt standard fission weapon has been used since this is believed to be typical of the Lance warhead that the new ERW will replace. In some situations a 1-kt standard fission weapon will be examined since this is more representative of the existing 203-mm shell.

In figure 6.3 the data on the effects of these weapons on tanks are directly compared. The old standard fission warhead for the 203-mm shell is compared in table 6.1 with the 1-kt enhanced radiation version.

From these data it will be seen that the enhanced radiation 1-kt Lance warhead will have approximately the same radiation effects on tank crews as the existing 10-kt fission warhead for that missile. The radius at which tank crews will be put immediately and permanently out of action (8 000 rads) is significantly larger for both of these warheads than the

Figure 6.3. Effectiveness of fission and enhanced radiation weapons against tanks

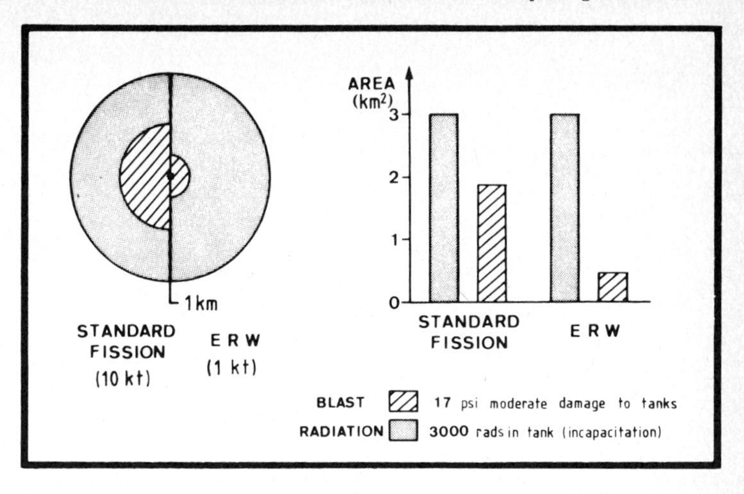

Table 6.1. Radii of weapon effects (metres)

Weapon	Burst height (metres)	Radiation dose			Blast effects		
		8 000 rads	3 000 rads	650 rads	17 psi	6 psi	3 psi
1-kt ER	150	690	820	1 100	280	430	760
1-kt fission	150	360	440	690	300	520	910
10-kt fission	150	690	820	1 100	640	910	1 520

Source: Except for 17-psi blast data, reference [3].

radius at which the tanks themselves will be damaged by blast. Thus from the point of view of radiation effects there is no difference between the new neutron warhead for the Lance and the existing fission one. The range of blast damage to tanks will be much greater for the old Lance warhead than for the new ERW (more than twice as great), and this could be militarily very significant.

In the case of the 203-mm artillery shell, which will have a 1-kt yield for both the enhanced radiation and the existing fission warheads, the ERW will incapacitate tank crews at almost twice the distance. The range of blast effects on tanks will be slightly greater for the standard fission weapon, but in both cases quite small and not sufficient to justify the use of nuclear weapons.

Kent Wisner [4] has concluded that the actual military advantage of the greater range of neutron effects of the ERW warhead in the 203-mm shell is quite limited in light of Soviet doctrine for tank attacks. This analysis shows that only when a Soviet tank battalion approaches the line of contact with enemy forces (4–6 km away) and deploys into complete columns (750–1 000 m apart) will there be significant advantage for the ERWs. At greater distances and on closer contact the expenditure of nuclear munitions will be the same regardless of the type of warhead. Thus only under limited circumstances will the enhanced radiation warhead for the 203-mm shell provide significantly greater military capability for destroying tank crews by radiation.

But even this limited advantage of the ERWs is not realistically useful for repelling tank attacks on the battlefield. Killing the tank crews is not a very satisfactory way of stopping tanks because a battlefield commander can never be sure when the enemy tank has been put out of action. Unless a very high super-lethal exposure is obtained, the tank crew might be able to continue fighting even though it was doomed eventually to die. Even if the crew had been put out of action there would be no way to be certain that this was the case by external observation. There would always be the fear that the crew would temporarily recover and be able to continue combat as a kamikaze unit. Since the tank itself would be undamaged at the distance that the crew had been killed, there would also be the possibility that new crews could take over the tank.

On the battlefield the most satisfactory way of knocking out a tank column is to destroy visibly the tank itself. This can only be done by blast from a nuclear weapon or by hitting the tank using conventional PGMs. Under these circumstances there is no question that the tank has been put out of action. Unfortunately neither the enhanced radiation nor the standard fission 1-kt warheads for the 203-mm shell will do this at significant distances. However, the existing 10-kt fission warhead for the Lance will destroy tanks to a distance of about 0.5 km. Thus if nuclear weapons were used to repel a Soviet tank attack, there is little doubt that the existing Lance warhead is superior to the ERW version [5].

Since neither 203-mm artillery shell is particularly effective, PGMs would seem to offer a far better and less dangerous alternative. Even in the case of the Lance, serious questions can be raised as to whether the increased kill radius from blast of the nuclear alternative is superior to reliance on conventional munitions. If the threshold between conventional and nuclear weapons is to be crossed, certainly the military effectiveness of the nuclear round must be very much greater to make such escalation truly necessary.

VI. Collateral effects

Blast damage

The major differences between enhanced radiation warheads and the existing fission warheads will be in the collateral blast damage to urban structures. The ranges of these effects are summarized in table 6.1 and figure 6.4. Collateral damage will be relatively low but by no means completely absent for the 1-kt warheads for either of the 203-mm shells and for the enhanced radiation Lance warhead. Were the tanks to be attacked in a village, all the buildings within about 500 m would be severely damaged. The Hiroshima bomb had a range of blast damage less than $2\frac{1}{2}$ times that of the 1-kt warheads. For the existing 10-kt standard fission warhead, almost equivalent to the Hiroshima bomb, the blast damage can be very large, the damage radius being more than twice as great as for the 1-kt ERW. The primary argument for replacing the current Lance warhead with the lower-yield enhanced radiation version was to reduce these collateral blast effects. This reduction would make the decision for the first use of nuclear weapons easier since the field commander—and ultimately the President of the United States—would not be in the position of ordering a defence that would destroy the cities of Western Europe; but realistically these smaller weapons will not spare

Figure 6.4. Collateral effects of fission and enhanced radiation weapons

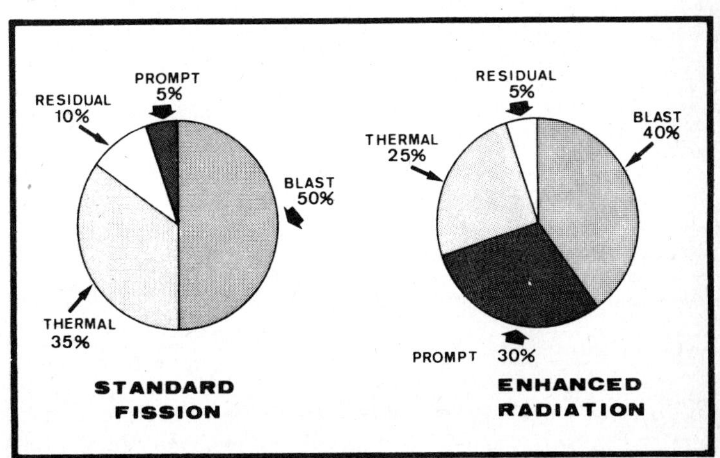

European cities. Once nuclear weapons have been used on the battlefields in Europe, whether they be enhanced radiation or fission weapons, the Soviet Union would almost certainly respond with its most effective warheads and disregard their effect on neighbouring buildings. US reliance on ERWs will not ensure the survival of European cities in the event of a nuclear war. The only way to avoid such widespread destruction would be to rely on conventional PGMs.

Radiation exposure of civilian and military bystanders

The damage to buildings is not the only criterion that will determine the acceptability of nuclear explosions in the inhabited areas of Western Europe. The radiation exposure of the population is also very important. Civilian bystanders, just like tank crews, are susceptible to becoming casualties from nuclear radiation. Friendly troops must also be spared the long-term radiation effects. Even if they survived initially, it would not be acceptable for military men to die miserably years after the conflict was over. This means that nuclear weapons cannot be used close to friendly forces, a serious drawback in a mobile, often confused battlefield situation.

A 1-kt enhanced radiation warhead would produce an exposure of 150 rads at about 1.5 km from the point of detonation (see figure 6.4). Even if people were in shelters, reducing the exposure by a factor of five, they would be liable to suffer long-term radiation effects, perhaps leukaemia, eye cataracts or genetic damage. Although the effects are not so precisely known, neutrons are more prone to produce long-term biological effects than is gamma radiation. Significant radiation casualties could be expected over an area of 10 km^2 for each enhanced radiation weapon used. Since any major conflict in Europe in which neutron weapons were used to repulse Soviet tank thrusts could involve 1 000 to 10 000 such battlefield weapons, the total area in which Europeans could be subjected to dangerous neutron exposures could range from 10 000 to 100 000 km^2. The number of actual casualties would, of course, depend upon the exact area in which the tank battles occur, but the number will almost certainly be very large. The neutron bomb is not a prescription for a safe nuclear war for Europeans.

Radioactive contamination

The deposition of radioactive materials around the battle zone or on civilian populations can be a serious problem for both ensuing military operations and the safety of civilian bystanders. Neither the standard fission weapons nor the neutron bombs will produce serious fall-out of radioactive fission products on the battlefield in the vicinity of the point of the explosion provided that the height of burst is high enough to prevent the fireball from touching the ground (about 150 m for a 10-kt explosion, or 75 m for a 1-kt explosion). Furthermore this height of burst is about the optimum for putting tanks out of action (17 psi) and is normally desirable unless hardened underground shelters are being attacked. Therefore, even if local fission product fall-out is to be avoided, there is little advantage in having a weapon deriving only a small fraction of its energy from fission. On the other hand, the neutrons from ERWs react with materials in the soil to produce induced radioactivity in a circle around 'ground zero' (the point on the ground directly beneath the point of detonation). This induced radioactivity will be greater than for fission weapons of the same yield. It would be sufficient to prevent unlimited occupation of an area of 3 km^2 around ground zero for a couple of days with an ERW of 1 kt exploded at a height of 75 m. This could be an important drawback to the use of ERWs in the event that friendly forces wish to occupy the area after the tank attack had been repulsed.

VII. Likelihood of nuclear war

No one can predict how the availability of ERWs will influence the decision to initiate the use of nuclear weapons in a European conflict. The precise nature of the conflict, the attitudes of the leaders and military commanders of the time and the perceptions of the risks at that time will all have major impacts. Proponents of these weapons make the claim that the decision to employ the nuclear wearheads will be easier in view of their reduced blast destruction in urban areas. Thus they argue that the deterrent to a Soviet tank attack will be much more credible. But if the decision to use these weapons is really easier, it should also increase the chances that a conflict will become nuclear. Proponents assert that, nevertheless, nuclear war will not be any more likely, because any President of the United States would recognize the significance of the first use of nuclear weapons and would not be influenced to make such a decision just because of the reduced blast effects. President Carter made a statement to this effect when he authorized proceeding with the programme in 1978. If the USSR recognizes this, then there is no basis for enhanced deterrence. This dichotomy between credibility of the deterrent and likelihood of use can never be resolved in advance of an actual confrontation. The risks that a nuclear war presents to our civilization certainly demand that no steps be taken that would increase the chances that nuclear weapons are actually used in any conflict.

VIII. Conclusions

The military advantages of the neutron bomb over existing nuclear weapons have been greatly exaggerated by political leaders in the West. The increased dangers they present to the peoples of Western Europe in comparison with existing nuclear warheads have been greatly exaggerated by political leaders in the East. The use of any nuclear weapons, fission or neutron bombs, would be an unparalleled disaster for Europe.

The Lance enhanced radiation warheads are probably less effective in repelling tank attacks than the fission warheads they replace, and they will reduce urban blast damage only if the Soviet Union does not retaliate with fission weapons of its own. The 203-mm ERW shells have an enhanced capability for putting tank crews out of action in limited combat situations, but have essentially the same collateral effects as the current fission versions. Neither the fission nor the ERW warheads for the 203-mm shell are particularly effective in destroying tanks themselves and do not warrant their first use in place of PGMs.

Crossing the threshold from the use of conventional weapons to the first use of any nuclear weapons will create an extremely high risk of escalation to all-out nuclear war. The contribution of the deployment of neutron bombs in Europe to the deterrence of Soviet aggression would appear marginal, and it could make it easier to cross this threshold and thus make the devastation in Europe and probably the world more likely.

The neutron bomb is the wrong approach to the modernization of nuclear weapons in Europe. The aim should be to reduce, not increase, the likelihood of their use. They serve only as the ultimate deterrent to Soviet aggression, for if they are ever used they will have failed in their purpose. The West must move to decrease reliance on any nuclear weapons to meet military requirements and move to a position where Western conventional weapons can deter any conventional attack from the East.

References

1. Scoville, Herbert, Jr., *A Comparison of the Effects of 'Neutron Bombs' and Standard Fission Weapons*, Center for Defense Information Conference Report, 24 April 1981.
2. SIPRI, *Nuclear Radiation in Warfare* (Taylor & Francis, London, 1981).
3. Cohen, S. T., *The Neutron Bomb: Political, Technological and Military Issues* (Institute for Foreign Policy Analysis, Cambridge, Mass., November 1978).
4. Wisner, K., 'Military aspects of enhanced radiation weapons', *Survival*, November/December 1981.
5. 'Soviet armour and the "neutron bomb" ', *Defence Attaché*, January–February 1979.

7. Nuclear explosions

Square-bracketed numbers, thus [1], *refer to the list of references on page* 133.

As many as 1 321 nuclear explosions were conducted during the period from 1945 to 1981. The USA and the USSR are responsible for conducting over 87 per cent of all these explosions.

I. Explosions in 1981

Of the 49 nuclear explosions which took place in 1981, the USSR carried out 21. (Five of these were conducted outside the known Soviet weapon testing sites and are therefore presumed to have served non-weapon purposes.) The USA conducted 16 nuclear weapon test explosions in the usual site in Nevada; the UK conducted 1, also in Nevada; and France conducted 11 on the atoll of Mururoa in the Pacific Ocean. China did not test at all last year.

All explosions in 1981 were carried out underground and, according to data obtained from the Hagfors Observatory in Sweden, all had a yield below or around 150 kt (the yield of the French tests was 20 kt or below).

Since 1978 the rate of nuclear testing has remained at a level of about one per week.

In recent years there have been reports that the atoll where the French tests are conducted was severely damaged by explosions and that, as a result, radiation was leaking into the Pacific Ocean. It has been revealed that in 1979 a nuclear explosive device stuck half-way down the test shaft so that when it was fired, the explosion split the rock through to the sea and caused a tidal wave which damaged certain installations [1]. In addition to the complaints about these occurrences made by a French trade union and the protests by international ecological movements, there have also been official expressions of concern from Australia and New Zealand about possible nuclear pollution [2].

On 9 December 1981 the French Defence Minister, speaking in the French National Assembly, affirmed that the atoll of Mururoa was sinking due to natural processes rather than from repeated underground blasts. The Minister admitted, however, that on 11–12 March 1981 a storm had dispersed radioactive products from pre-1975 testing, contained under an asphalt surfacing. He added, without elaborating, that this had created "a new radiological situation" and that all the necessary precautionary measures had been taken [3].

II. Military significance of nuclear tests

Nuclear weapons are tested mainly in order to find ways of increasing their efficiency and to develop new weapon designs (particularly to improve yield-to-weight ratios); to study the effects of the blast, heat, radiation and fall-out produced by a nuclear explosion; to develop mechanisms to ensure the safety and security of nuclear devices; and to maintain confidence in the reliability of stockpiled weapons.

However, many scientists claim that nuclear weapon technology has reached a 'state of maturity': while further development may lead to some increases in the efficiency of the weapon or its adaptation to specialized missions, it is not likely to result in qualitatively new developments [4]. In fact, despite predictions which have been made, no significant breakthroughs have taken place in this field for the past 20 years or more. Even the 'neutron bomb', a controversial political issue in recent years, is actually an invention of the late 1950s and early 1960s [5a]. Improvements in the performance of both strategic and tactical nuclear weapon systems are more the result of the evolution of the non-nuclear components, in particular the development of new delivery vehicles, than of improved designs of the nuclear explosive component. In other words, it seems that new military requirements could be met by previously tested, off-the-shelf designs.

Concerning the weapon effects, the more than 1 300 tests carried out during the past 36 years ought to have provided ample information. There are, of course, many uncertainties regarding a nuclear war, but those related to the physics of weapons are not fundamental and further testing to remove them is deemed to be of relatively little value [6].

To enhance the safety and security of nuclear devices, so-called permissive action links to prevent the use of weapons by unauthorized personnel, as well as use-denial mechanisms, which disable the nuclear warhead to prevent its use by terrorists, are being deployed. Moreover, insensitive high explosives which are resistant to crashes, fire or bullets are being developed [7]. Other improvements of this kind can certainly be made, but much can be accomplished without experimental explosions [5b].

It is further contended that test explosions are needed to check the performance of existing weapons and to correct possible defects. However, there exists authoritative evidence that the continued operability of stockpiled nuclear weapons can be achieved by non-nuclear testing [8]. As a matter of fact, very few reliability tests have been conducted, at least in the USA. Even if nuclear weapons actually were subject to degradation in the arsenals, this could be considered a gain for the cause of arms control rather than a loss for international security, on the condition that the parties were equally affected.

Should confidence in the reliability of weapons diminish, this is more likely to influence those planning for first use than those planning for retaliation only. The effect, if any, would be to widen the fire-break between conventional and nuclear weapons and to shift the role of nuclear weapons gradually towards that of weapons useful only for the deterrence of nuclear attack [9].

III. Existing limitations on nuclear testing

The question of stopping nuclear weapon tests has been on the agenda of multilateral, bilateral (US–Soviet) and trilateral (UK–US–Soviet) negotiations ever since the early 1950s. More than 40 UN General Assembly resolutions dealing with this subject have been adopted over the past years, and on seven occasions the General Assembly condemned all nuclear tests in the strongest terms. Nevertheless, no comprehensive ban had been reached by 1982. The 1963 Partial Test Ban Treaty (PTBT), the 1974 Threshold Test Ban Treaty (TTBT) and the 1976 Peaceful Nuclear Explosions Treaty (PNET) (see chapter 13 for summaries of the provisions of the treaties)—the three partial agreements which have been signed so

Figure 7.1. Nuclear explosions conducted before and after the PTBT

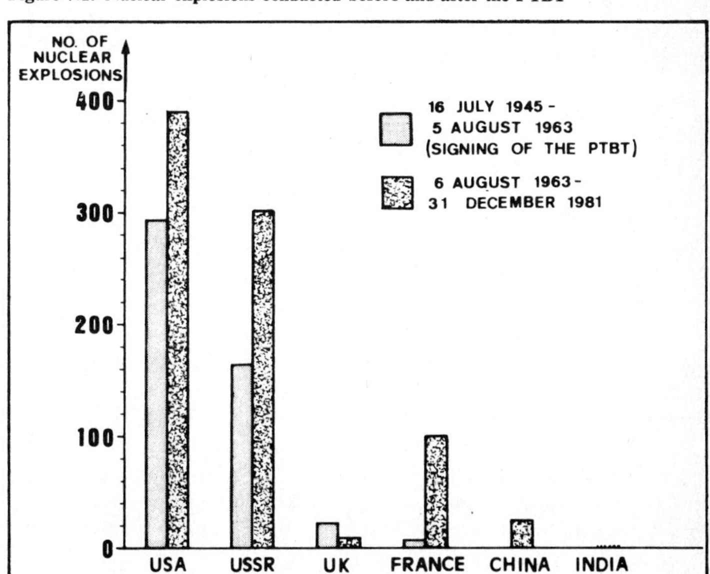

far—have only circumscribed the environment for nuclear testing and reduced the size of the explosions. After the signing of the PTBT, for example, the USA and the USSR have carried out a considerably greater number of explosions than before (see figure 7.1). They have been able therefore to develop new generations of nuclear warheads and delivery vehicles, and the nuclear arms race has continued unhampered. The commitment under the TTBT to restrict the number of tests to a minimum has had little or no effect on the testing activities of the major powers, and the yield threshold established by this Treaty—150 kt—is so high (more than 10 times higher than the yield of the Hiroshima bomb) that the parties cannot be experiencing onerous restraint in continuing their nuclear weapon programmes. Finally, the PNET restrictions have only provided an indispensable complement to the TTBT to ensure that peaceful nuclear explosions should not provide weapon-related information that is not obtainable from limited weapon testing.

IV. Negotiations for a comprehensive test ban

In 1977 the UK, the USA and the USSR engaged in trilateral talks for the achievement of a comprehensive test ban (CTB), but in 1980, with the change of the US Administration, the talks were adjourned *sine die*. Up to that time a certain measure of agreement was reached among the negotiators [10].

Main points of agreement

The UK, the USA and the USSR agreed that: (*a*) a comprehensive test ban treaty should prohibit any nuclear weapon test explosion in any environment and be accompanied by a protocol on nuclear explosions for peaceful purposes, which would establish a moratorium on such explosions; (*b*) any amendment to the treaty would require the approval of a majority of parties, which majority should include all parties that are permanent members of the UN Security Council, and a conference would be held at an appropriate time to review the operation of the treaty; (*c*) the parties would use national technical means of verification at their disposal to verify compliance and would undertake not to interfere with such means of verification; an international exchange of seismic data would be established; and (*d*) the treaty would provide for consultations to resolve questions that may arise concerning compliance and any party would have the right to request an on-site inspection for the purpose of ascertaining whether or not an event on the territory of another party was a nuclear explosion.

While verification no longer seems to be a major obstacle, a series of complex technical problems related to verification remains to be solved.

Verification problems

Whatever additional methods might be used, seismological means of verification will certainly constitute the principal component of an international control system for an underground test ban. With this in mind, the Geneva-based Committee on Disarmament established an *ad hoc* group of scientific experts to consider international co-operative measures to detect and identify seismic events. The group has suggested that these measures should include a systematic improvement of procedures at seismological observatories around the globe, an international exchange of seismic data and the processing of the data at special international data centres.

In particular, the *ad hoc* group of experts considers that a seismological verification system should comprise about 50 globally distributed teleseismic stations selected in accordance with seismological requirements. These would be national facilities operated in accordance with generally accepted rules. The seismograph stations belonging to the system would routinely report the parameters of detected seismic signals, as well as transmit data in response to requests for additional information regarding events of particular interest. International centres would receive the data mentioned above; apply agreed analysis procedures to these data in order to estimate location, magnitude and depth of seismic events; associate identification parameters with these events; distribute compilations of the complete results of these analyses; and act as a data bank [11].

Although the global seismic network can provide a high degree of confidence that a comprehensive test ban is not being violated, there may still be events of uncertain origin. One way to reduce this uncertainty, which in most cases will be related to earthquake areas, could be for the state in question to provide seismic data for the suspected event from local stations not belonging to the global network.

The UK, the USA and the USSR agreed to develop measures of reciprocal verification, independent of the envisaged international co-operative measures, in order to obtain supplemental seismic data from high-quality, tamper-proof national seismic stations (NSSs) of agreed characteristics. Ten NSSs would be installed on the territories of the USA and of the USSR, but no agreement could be reached regarding the number of such stations in the UK. Questions regarding the specific locations of the NSSs, their emplacement and maintenance as well as the transmission of data produced by them have not been settled.

While the three negotiating powers agreed on the possibility of having on-site inspections, the procedure for setting in motion the inspection process (including the nature of the evidence needed to justify a request for on-site inspection), the modalities of the inspection itself (including the equipment to be used), as well as the number, rights and functions of the inspectors, have yet to be specified.

Other unresolved issues

Among other issues which remain to be settled is the status of laboratory tests which could, for example, consist of extremely low-yield nuclear experiments or the so-called inertial confinement fusion [12].

Extremely low-yield nuclear experiments could involve an explosion of a device which may have the same characteristics as a nuclear explosive device but which uses fissile material of an amount or kind that produces only a fraction of the yield of the chemical explosion that sets off the release of the nuclear energy. The question is whether such a test, which could be conducted in a laboratory, should be considered a nuclear weapon test explosion. The inertial confinement concept is to use lasers or other high-power sources to heat and compress small pellets containing fusionable fuel (deuterium and tritium). If a properly shaped pulse of sufficient energy can be delivered to the pellet, the density and temperature may become high enough for fusion. This would be a laboratory nuclear explosion of tiny proportions.

It may be argued that, in order to be effective, a comprehensive test ban should cover all explosions without exception, including laboratory tests. On the other hand, it can be contended that a comprehensive test ban could not cover laboratory tests because they are contained and not verifiable, and also because some of them may be useful for various peaceful purposes, including the development of new sources of energy.

Yet another point at issue is the duration of a comprehensive test ban treaty. The treaty negotiated trilaterally was planned to have a duration of no more than three years. The USA did not want to make a provision for a possible extension of the ban, while the USSR preferred to stipulate that the ban would continue unless the other nuclear weapon powers, not party to the treaty, continued testing. A ban of fixed duration would not fulfil the pledge included in the PTBT to achieve the discontinuance of all test explosions of nuclear weapons for all time. Moreover, a treaty of short duration would create a problem with respect to the adherence of non-nuclear weapon states, particularly parties to the NPT, which have renounced the possession of nuclear explosive devices for a much longer period. Finally, resumption of tests upon the expiration of a short-lived comprehensive test ban treaty would probably hurt the cause of arms limitation and disarmament more than if the treaty had never been entered into.

V. Conclusions

The discontinuation of nuclear weapon test explosions would not stop all improvements in nuclear warheads; certain improvements do not require tests involving nuclear reactions. A CTB would, nevertheless, have an arms limitation impact in that it would make it difficult, if not impossible, for the nuclear weapon parties to develop new weapon designs and would place constraints on the modification of existing designs. It would thereby narrow one channel of arms competition among the major powers. The arms control benefits could be further enhanced if the CTBT were followed by a ban on the production of fissionable material for weapon purposes. Such a 'cut-off' would slow the manufacture of nuclear weapons and could perhaps even be a step towards eventually ending this manufacture.

A CTB would also reinforce the Non-Proliferation Treaty by demonstrating the major powers' awareness of their legal obligation to bring the nuclear arms race to a halt. On the other hand, it is not certain that it would actually hinder the further proliferation of nuclear weapons, since a test explosion may not be absolutely essential for constructing at least a simple fission device. Nor is it certain that a CTB would provide sufficient incentives for the present non-NPT states to join the NPT, especially if these states have kept their nuclear weapon option open irrespective of the behaviour of the great powers, or if they consider that the mere cessation of tests by the nuclear weapon states is not a sufficient *quid pro quo* for their renunciation of nuclear weapons.

References

1. *Le Point*, No. 482, 14 December 1981.
2. *International Herald Tribune*, 24–25 December 1981.
3. *Le Monde*, 11 December 1981.
4. F.A.S. Public Interest Report Special Issue, *Comprehensive Test Ban*, Vol. 31, No. 6, June 1978.
5. York, H. and Greb, G.A., *The Comprehensive Nuclear Test Ban*, Discussion Paper No. 84, California Seminar on Arms Control and Foreign Policy, Santa Monica, California, June 1979.
   (a) —, p. 19.
   (b) —, p. 22.
6. Wiesner, J. B. and York, H., 'National security and the nuclear test ban', *Scientific American*, No. 211, October 1964, pp. 30–31.
7. Intelligence and Military Application of Nuclear Energy Sub-Committee of the Committee on Armed Services, *Current Negotiations on the Comprehensive Test Ban Treaty*, House of Representatives, 95th Congress, 2nd Session (US Government Printing Office, Washington, D.C., 1978), p. 73.
8. *Effects of a Comprehensive Test Ban Treaty on United States National Security Interests*, Hearings before the Panel on the strategic arms limitation talks and the

comprehensive test ban treaty of the Intelligence and Military Application of Nuclear Energy Subcommittee of the Committee on Armed Services, House of Representatives, 95th Congress, 2nd Session, August 14, 15, 1978, H.A.S.C. No. 95–89 (US Government Printing Office, Washington, D.C., 1978), pp. 133 and 181.

9. SIPRI, *The Test Ban*, SIPRI Research Report (Stockholm International Peace Research Institute, October 1971).
10. Committee on Disarmament document CD/130, 30 July 1980.
11. Conference of the Committee on Disarmament document CCD/558, 9 March 1978.
12. *Comprehensive Nuclear Test Ban*, Report of the UN Secretary-General, Committee on Disarmament document CD/86, 16 April 1980, pp. 38–39.

Appendix 7A

Nuclear explosions, 1945–81 (known and presumed)

I. 16 July 1945–5 August 1963 (the signing of the Partial Test Ban Treaty)

USA	USSR	UK	France	Total
293	164	23	8	**488**

II. 6 August 1963–31 December 1981

a atmospheric
u underground

	USA		USSR		UK		France		China		India		
Year	a	u	a	u	a	u	a	u	a	u	a	u	Total
6 Aug–31 Dec 1963	0	14	0	0	0	0	0	1					**15**
1964	0	28	0	6	0	1	0	3	1	0			**39**
1965	0	29	0	9	0	1	0	4	1	0			**44**
1966	0	40	0	15	0	0	5	1	3	0			**64**
1967	0	29	0	15	0	0	3	0	2	0			**49**
1968	0	39[a]	0	13	0	0	5	0	1	0			**58**
1969	0	28	0	15	0	0	0	0	1	1			**45**
1970	0	33	0	12	0	0	8	0	1	0			**54**
1971	0	15	0	19	0	0	5	0	1	0			**40**
1972	0	15	0	22	0	0	3	0	2	0			**42**
1973	0	11	0	14	0	0	5	0	1	0			**31**
1974	0	9	0	19	0	1	7	0	1	0	0	1	**38**
1975	0	16	0	15	0	0	0	2	0	1	0	0	**34**
1976	0	15	0	17	0	1	0	4	3	1	0	0	**41**
1977	0	12	0	16	0	0	0	6	1	0	0	0	**35**
1978	0	12	0	27	0	2	0	7	2	1	0	0	**51**
1979	0	15	0	29	0	1	0	9	0	0	0	0	**54**
1980	0	14	0	21	0	3	0	11	1	0	0	0	**50**
1981	0	16	0	21	0	1	0	11	0	0	0	0	**49[b]**
Total	**0**	**390**	**0**	**305**	**0**	**11**	**41**	**59**	**22**	**4**	**0**	**1**	**833**

III. 16 July 1945–31 December 1981

USA	USSR	UK	France	China	India	Total
683	469	34	108	26	1	**1 321**

[a] Five devices used simultaneously in the same test are counted here as one.
[b] The data for 1981 are preliminary.

8. Long-range theatre nuclear forces in Europe

Square-bracketed numbers, thus [1], *refer to the list of references on page* 181.

I. The issues

The current debate on nuclear weapons in Europe may significantly affect the political orientation of European countries and the shape of their defences. It has already had a substantial impact on European threat perceptions: while the two major powers perceive each other as posing the gravest threat to their security, many Europeans see the US–Soviet conflict as representing the gravest threat to European security.

At the centre of attention are the long-range theatre nuclear forces (LRTNFs). The essential aspects of the LRTNF issue may be summarized as follows.

1. New Soviet weapons, in particular the SS-20 missile, have provoked considerable concern in Western Europe. In times of peace, they are a source of anxiety; in times of crisis, they could be used for purposes of intimidation and blackmail; in times of war, Soviet doctrine emphasizes initiative, surprise, deep strikes and massive use, which can now be executed with greater precision than before. The Soviet Union has a numerical lead of more than 2:1 in LRTN systems—in aircraft as well as missiles—within striking range of Europe.

2. The military rationale for the NATO decision to deploy cruise and Pershing missiles was to keep open the option of striking a substantial number of targets in the USSR from Western Europe (thereby enhancing the 'nuclear umbrella' over Western Europe). Existing forces were no longer considered adequate for that purpose. Politically, however, the need for new weapons was ascribed to the Soviet LRTNF build-up—initially by West European politicians in particular. To justify the modernization request, the SS-20 was singled out for particular attention. However, Western Europe had been living under the shadow of Soviet LRTN missiles for almost 20 years.

3. For the Soviet Union, the replacement of old SS-4 and SS-5 missiles was technologically overdue, and the decision to deploy the SS-20 may have been taken without much consideration for its impact on international affairs. The concerns of leaders in the East and the West were therefore badly synchronized: while many Western politicians 'rediscovered' the Soviet missile threat when the SS-20 was introduced, Soviet leaders did nothing to allay the fears. For more than two years after the first missiles were deployed and for many months after NATO's

deployment plan was substantiated, the Soviet Union made no major political move on LRTNFs.

4. The US nuclear umbrella—the notion that in defence of Western Europe the United States is willing to use nuclear weapons not only on the battlefield, but also against the Soviet Union—has folded up. Should a war break out between the military alliances in Europe, both the USA and the USSR would do their utmost to keep their own territories out of the conflict. The deployment of cruise and Pershing missiles in Western Europe does not change this. If Soviet territory were struck by US nuclear weapons, it must be assumed that the Soviet Union would retaliate against US territory, regardless of the launching point or physical characteristics of the delivery vehicle. The targeting policy for cruise and Pershing missiles is therefore likely to comprise alternative sets of targets, tailored according to different war scenarios. In the case of a European battle, the missiles may be used against East European countries; in a strategic exchange, against the Soviet Union.

5. For the United States, the main military interest in deploying new missiles seems to be of a strategic nature. The Pershing II will be one of the most capable counterforce weapons in the US arsenal, should it ever be deployed in Europe. It has superior characteristics for limited strikes, is ideal for use against time-urgent targets (such as missiles, command and control centres, quick-reaction alert aircraft and submarines in port), and therefore fits the requirements of the countervailing strategy, codified in Presidential Directive 59.

6. For the European countries, new missiles make a difficult situation even worse. More effective war-fighting weapons, introduced in a major power competition which is not of European making but in which Europeans—East and West—may become the main losers, are clearly detrimental to their security. The host countries would, moreover, be burdened with a number of high-priority nuclear weapon targets, which would make it virtually certain that Western Europe would be drawn into any strategic war between the two great powers.

7. To avoid circumvention and ensure substantial limitations, the LRTNF negotiations that started in Geneva on 30 November 1981 should soon overlap with resumed US–Soviet talks on strategic arms, and also lead on to limitations of systems of shorter range. All targets that can be struck by the new Soviet and US theatre systems can, for instance, be hit by intercontinental systems as well. A mere reduction of LRTNFs will therefore lose much of its military significance if intercontinental systems are allowed to increase unchecked.

8. US LRTNFs in Western Europe can reach the Soviet Union, while the MIRVed SS-20 cannot reach the United States. Between the two major powers, parity in intercontinental systems and parity in LRTNFs

are therefore incompatible with overall strategic parity. Given that the SALT agreements have established a kind of parity in intercontinental systems, the only regional level which is compatible with overall strategic parity is that which is defined by the figure zero. A solution making the deployment of cruise and Pershing missiles in Western Europe superfluous would, therefore, not only enhance the security of European states, but also facilitate progress in US–Soviet strategic arms limitation.

9. Cruise missiles, the sea-launched version (SLCM) in particular, may become formidable obstacles to effective arms limitation. In an international atmosphere of deep distrust, they raise unprecedented demands for ingenuity in the field of verification; should current plans for wide dispersal of SLCMs be implemented, effective verification would become extremely difficult. Substantial limitations on this technology may therefore be of fundamental significance for the future of arms control. Zero-level agreements, prohibiting certain categories of weapon altogether, are by far the easiest to verify. Should the deployment of SLCMs in waters adjacent to Europe proceed and not be regulated within the framework of resumed talks on strategic arms, it would radically alter the data base for the Geneva LRTNF negotiations.

10. By the end of 1981, the Soviet Union had some 175 SS-20 missile launchers within striking range of Europe. On the assumption that each missile carries three MIRVs (multiple independently targetable re-entry vehicles), the total number of warheads equals that which was deployed on SS-4s and SS-5s before the SS-20 became operational. In terms of launchers, the present SS-20 arsenal is roughly equal to those of the UK and France combined (175 versus 162 launchers). In quantitative respects, the *status quo ante* and a matching of Soviet missile systems with those of the UK and France can therefore be obtained by eliminating all SS-4s and SS-5s and freezing the number of SS-20 launchers. Such a move, establishing a balance in the number of LRTN launchers in the region but without affecting the strategic balance between the two major powers, would facilitate further endeavours towards nuclear disarmament.

II. The history of LRTNFs

Definitions

Theatre nuclear weapons can be divided into three categories, according to range.

Long-range theatre nuclear forces (LRTNFs) are nuclear weapons with a maximum range of more than 1 000 km, but less than 5 500 km (intercontinental range). For many weapon systems, the range specification is somewhat arbitrary, but serves the purpose of focusing attention on a certain set of nuclear weapon systems. Nor is it easy to classify all systems according to this criterion: for instance, the Soviet SS-22 missile, the successor to the SS-12 Scaleboard, is accredited with a range of about 1 000 km—perhaps a little more or less—and a number of aircraft are also extremely difficult to classify. In Soviet terminology, LRTNFs are described as operational-strategic weapons and are allocated to the Strategic Rocket Forces.

Medium-range theatre nuclear forces (MRTNFs) have a range of 200 to 1 000 km. These weapons are designed to support operations at the corps-army level or, in the Soviet case, at the army-front level.

Short-range theatre nuclear forces (SRTNFs) have a range up to 200 km. Often designated 'battlefield' nuclear weapons, these are primarily intended for use at the division and corps levels.

The term 'LRTNFs' is often used interchangeably with 'eurostrategic weapons'; the term 'eurostrategic' refers to strategic uses against targets in Europe. 'Strategic' use refers to strikes against the socio-economic structure of the opponent, or his offensive and defensive strategic armoury and associated infrastructure. 'Tactical' use refers to attacks on targets with more or less direct effects on the course of battle. This dichotomy leaves a grey area of targets whose importance for the tactical situation is more remote, such as ports, roads, railway-yards, and command, control, communications and intelligence (C^3I) centres: LRTNFs can also be used for interdiction strikes against such targets (see section V). In the nuclear arms limitation talks that started in Geneva in November 1981, world-wide as well as regional, European limitations have been proposed; also for that reason, LRTNF is the more appropriate term to use.

The US forward based systems

In the summer of 1949, the United States deployed 32 B-29 bombers in the UK. The B-29 'superfortress' had a radius of operation of about 2 500 km, and therefore depended on forward bases for strikes against the Soviet Union. This was the beginning of the US forward based systems (FBSs) in Europe.[1]

At this time, the B-52 was on the drawing boards. However, in order to acquire jet-bomber capability as soon as possible, priority was given to the Boeing B-47 medium-range bombers; the technological challenge was less than for an intercontinental aircraft, and the overseas bases

[1] In November 1946, six B-29s 'toured' Europe and surveyed airfields for possible use. This is regarded as the first instance in which SAC bombers were used as an instrument of international diplomacy [1].

were regarded as safe. The B-47 entered operational service in 1951, and remained the mainstay of the US Strategic Air Command (SAC) for 10 years. More than 2 000 were built, and the last ones were phased out in 1966. B-47s operated from bases in French Morocco, Spain and the UK, with units rotating from the continental United States (CONUS).

Throughout the 1950s, a variety of other nuclear-capable aircraft—both land- and carrier-based—were also deployed in Europe and in European waters, some of them capable of striking against the Soviet Union.

The Karman Committee of 1945, which summarized the recent advances in science and technology, concluded that the USA should concentrate on developing jet aircraft, whereas missiles were relegated to the more distant future [2]. Nevertheless, the military services began small-scale missile programmes, often based on technology inherited from German wartime efforts. In the field of long-range vehicles, efforts were concentrated on aerodynamic, 'cruise' missiles. The Navy operated its dual-capable 650-km range Regulus cruise missile on board submarines from 1954 to 1964. The Air Force missile programme was somewhat more ambitious, and more than 1 000 dual-capable, supersonic Matador cruise missiles, with a range of about 800 km, were produced. The Matador was placed with units in the Federal Republic of Germany in the mid-1950s. Some years later, it was replaced by another cruise missile, the Mace A/B, with a range of up to 2 500 km. The Mace was withdrawn in the second half of the 1960s because of its vulnerability to new generations of jet-propelled air-defence aircraft.[2]

At the NATO meeting in Washington, D.C. in December 1957, it was decided to deploy long-range ballistic missiles in Europe. Around 1960, US Thor and Jupiter missiles became operational in the UK, Italy and Turkey. They had a range of approximately 3 000 km, and a warhead yield of 1.5 megatons (Mt). The Thor missiles deployed in the UK (60) were deactivated by the end of 1963, while the Jupiters (30 in Italy and 15 in Turkey) were phased out by 1965 [4]. The modest numbers and short lifetime were due to slow count-down, high vulnerability and, more importantly, the introduction of submarine-launched ballistic missiles (SLBMs) and intercontinental ballistic missiles (ICBMs). Polaris submarines were already patrolling the Mediterranean and the Norwegian Seas when the land-based missiles were withdrawn.

The advent of Soviet LRTNFs

Soviet LRTNF deployment came largely in response to the US forward based systems. They also compensated for the US lead in intercontinental

[2] At peak deployment there were five Mace A squadrons and one Mace B squadron in hardened sites in Europe, with 20–50 missiles per squadron [3].

weapons. While waiting for their own intercontinental missiles, the Soviet Union held Western Europe hostage. Finally, Soviet LRTNFs must be seen in relation to the British, French and, in Asia, to the Chinese nuclear forces capable of hitting Soviet territory.

Soviet LRTNFs reached a peak in the mid-1960s, when altogether 733 missiles were operational. Approximately 100 were directed against the Middle East, South Asia and the Western Pacific, and the rest were available for strikes against Western Europe, together with 880 bomber aircraft. The missiles were of three types: the 1 200-km range SS-3s (only 40), the 1 800-km range SS-4s, and the 3 500-km range SS-5s. All of Western Europe was within range of Soviet megaton-yield warheads. The bombers were of two types: the Tu-16 Badger and the Tu-22 Blinder.

While the SS-3 missiles were withdrawn, the increasingly vulnerable SS-4s and SS-5s were retained. Already by the mid-1960s the Soviet Union tried to resolve the vulnerability problem by developing a new mobile land-based missile, the SS-14 Scapegoat (designated Scamp when vehicle-mounted). However, it seems to have been a technological failure (although a small number of them were deployed in the Far East). Subsequently, intercontinental SS-11 missiles, and later also SS-19s, were deployed in the European theatre. At the same time, these deployments appeared to compensate for the transfer of part of the SS-4/SS-5 force to the Chinese border in 1968. At least 120 SS-11s and 60 SS-19s were deployed at SS-4/SS-5 sites at Derazhnya and Pervomaysk.[3] The mobile, intercontinental SS-16 missile, which was prohibited by SALT, finally gave rise to the SS-20, deployed from 1976/77 on: the SS-20 basically consists of the first two stages of the SS-16.

For intelligence services and military experts, the introduction of the SS-20 was therefore no surprise; on the contrary, it was technologically overdue. Moreover, theatre nuclear missiles had already been targeted on military–economic centres (such as ports and industrial centres), military and political command and control facilities, and strategic nuclear force components (such as airfields, nuclear weapon depots and detection and warning systems). So, while the SS-20 meant a leap forward in counterforce capability, it represented no radical departure in doctrine. Both technologically and doctrinally, the phasing in of SS-20s was a 'natural', almost unquestionable move. The decision may have seemed an easy one to make, and to a large extent it may have been reduced to a matter of military-bureaucratic automaticity, without much consideration of its impact on international affairs. However, for many Western political

[3] Garthoff indicates that the number of ICBMs designated for the European theatre has been in the range of 180–360 [5].

circles, the new missiles were seen as a sign of Soviet threat and aggressiveness. At a time of increasing East–West tension, exaggerations of the threat—both unintentional and deliberate—were only to be expected.

For more than two years after the first SS-20s were deployed, the Soviet Union neither took a major initiative nor made a major political statement on LRTNFs. When Brezhnev finally spoke in Berlin on 6 October 1979, he offered too little too late: too little, because the offer to reduce the number of launchers did not preclude an increase in the number of warheads targeted on Western Europe; and too late, because in effect, NATO's decision of 12 December 1979 had already been taken. Had the Soviet Union, for instance—as a follow-up to Brezhnev's visit to Bonn in June 1978, where LRTNFs figured prominently on the agenda—promised that it would not deploy more warheads on SS-20s than it would eliminate by removing old SS-4 and SS-5 missiles, much fuss might have been avoided. Then the search for a zero solution, making deployment of new missiles for Western Europe superfluous, could have had a much better start.

At that stage, however, the Soviet leaders do not seem to have been sufficiently geared to the political aspects and consequences of their SS-20 deployments. The concerns of leaders in the East and the West were, in other words, badly synchronized: while being a 'matter of course' for Soviet leaders, many Western politicians 'rediscovered' the Soviet missile threat when the SS-20 was introduced. In the West, the SS-20 was presented as a grave, new threat—erroneously so—while in the East, leaders displayed no political activity to allay the fears—a major blunder.

III. Force comparisons

Any comparison of NATO and WTO forces should, ideally, be dynamic and qualitative, based on assessments of survivability, penetrability, reliability, targeting options and employment doctrines, accuracy, exchange scenarios and the endurance of C^3I. However, attempts at quantifying these factors are bound to be arbitrary, and the whole exercise of very uncertain validity. The overviews given in tables 8.1–8.7 are therefore confined to relatively simple, quantitative force comparisons only. Missiles and aircraft are treated separately, although they are of course closely linked functionally. Air-to-surface missiles (ASMs) are treated together with the aircraft.

Missiles

Ballistic missile systems that have been assigned to European missions but are accounted for in the SALT II Treaty, notably Soviet SS-11s,

SS-19s and SS-N-5s on Hotel II-class submarines, and US Poseidon warheads allocated to SACEUR (the Commander of NATO forces in Europe) for targeting, are not included in the comparison. In the official US and Soviet LRTNF estimates presented shortly before the opening of the Geneva talks, neither party included them. At the low end of the range spectrum, the Soviet SS-12 Scaleboard and the Western Pershing IA, with ranges of 800 and 740 km, respectively, clearly fall into the MRTNF category. The Soviet SS-22 missile, intended to replace the SS-12, has a somewhat longer range, but probably does not exceed 1 000 km. US figures include 100 SS-12/SS-22s, while the Soviet Union claims that only 50 SS-12s exist and that the SS-22 is not yet operational.

Sea-based cruise missiles such as the Soviet SS-N-3 Shaddock and the SS-N-12 Sandbox may be employed in strategic land-attack roles. However, they have ranges below 1 000 km, are intended primarily for anti-ship use, and are therefore not counted.[4]

The WTO arsenal

The Soviet SS-4 Sandal and the SS-5 Skean are inaccurate, high-yield (1 Mt) weapons. They are liquid-fuelled and have very long reaction times. Some were deployed in silos, but most (some 80 per cent) were surface-mounted and reloadable. The SS-20, on the other hand, scores high on readiness, mobility, accuracy, firepower and range. It must, however, be fired from physically prepared positions. In addition to the MIRVed version (with three 150-kt warheads), there seems to be at least one single-RV (re-entry vehicle) version, achieving intercontinental range. While not the 'wonder weapon' some Western commentators claim it to be, the SS-20 undoubtedly represents an order-of-magnitude improvement in the Soviet capability to destroy time-urgent and semi-hard targets.

Towards the end of 1981, about 250 SS-20s were operational, in regiments of nine launchers and possibly with one reload missile per launcher. If this reload is of the single-RV intercontinental version, it may constitute a reserve force for use against the United States. One-third of the SS-20s are deployed in the Western and one-third in the Far Eastern USSR, with the last third in a swing position near the Urals. Single RVs with long range may have been preferred particularly for deployment in that area, reaching the peripheries of the Eurasian land mass from relatively invulnerable positions. The SS-20 is first and foremost a Eurasian weapon system.

The Soviet Union still operates 13 diesel-powered ballistic missile submarines of the Golf II-class, with probable deployment of six in the

[4] Some sources claim that the SS-N-12 may be given a range of 3 000 km or more, with transonic speeds rather than the usual Mach 2.5 [6].

Table 8.1. Long-range theatre nuclear missiles

Country	Missile designation	Year first deployed	Range (km)	CEP (m)	Warhead(s)	Inventory[a] A	Inventory[a] B	Programme status
USSR	SS-4 Sandal	1959	1 800	2 400	1 × Mt }	350	253	Phasing out
	SS-5 Skean	1961	3 500	1 200	1 × Mt }			Phasing out
	SS-20	1976/77	5 000	400	3 × 150-kt MIRV 1 × ?[b]	250	243	Production rate approximately 50 per year
	SS-N-5 Serb	1963	1 200	n.a.	1 × Mt	30	18	3 each on Golf II submarines, 6 of which have been deployed in the Baltic since 1976
USA	Pershing II	1983	1 800	40	1 × ? (low-kt)		0	108 launchers to be deployed by 1985
	GLCM	1983	2 500	50	1 × ?[c]		0	464 missiles to be deployed by 1988
UK	Polaris A-3	1967	4 600	800	3 × 200-kt MRV		64	On 4 SSBNs, being replaced by the 'Chevaline'-system, probably with 6 warheads (MRV), each of 50 kt
	Trident II D-5[d]	1990s	10 000	250	10 × 335-kt MIRV		0	Replacing the 'Polaris'/'Chevaline' system from the 1990s, probably with 64 launchers on 4 submarines
France	SSBS S-3	1980	3 000	n.a.	1 × 1-Mt		18	Conversion from S-2 to be completed by 1983
	MSBS M-20	1977	3 000	n.a.	1 × 1-Mt		80	On 5 SSBNs
	MSBS M-4	1985	4 000	n.a.	6 × 150-kt MRV		0	On the 6th SSBN; retrofit to be completed by 1989; total programme estimate: 96

[a] For the USA and the USSR, the official numbers are given. A: Figures released by the Department of State following President Reagan's speech at the National Press Club on 18 November 1981. B: Figures given by Leonid Brezhnev in *Der Spiegel*, 2 November 1981, and by Vadim Zagladin before the Fifth Pugwash Workshop on Nuclear Forces in Europe, Geneva, 11–13 December 1981.

Two-thirds of the SS-4s, SS-5s and SS-20s are estimated to be within striking range of Europe.

[b] Some SS-20 missiles are equipped with a single warhead and may therefore have intercontinental range.

[c] The W.84 warhead, with a low, selectable yield.

[d] The British government has not yet announced any decision regarding Trident I or Trident II (nor the number of submarines or missiles per submarine). Trident II seems the more likely because this missile will become the mainstay of the US SLBM force.

Range and yield are based on the likely US choice of warheads; since the UK will supply its own charges, it may choose other force specifications.

Figure 8.1. Deployments and maximum ranges of LRTN missiles in Europe

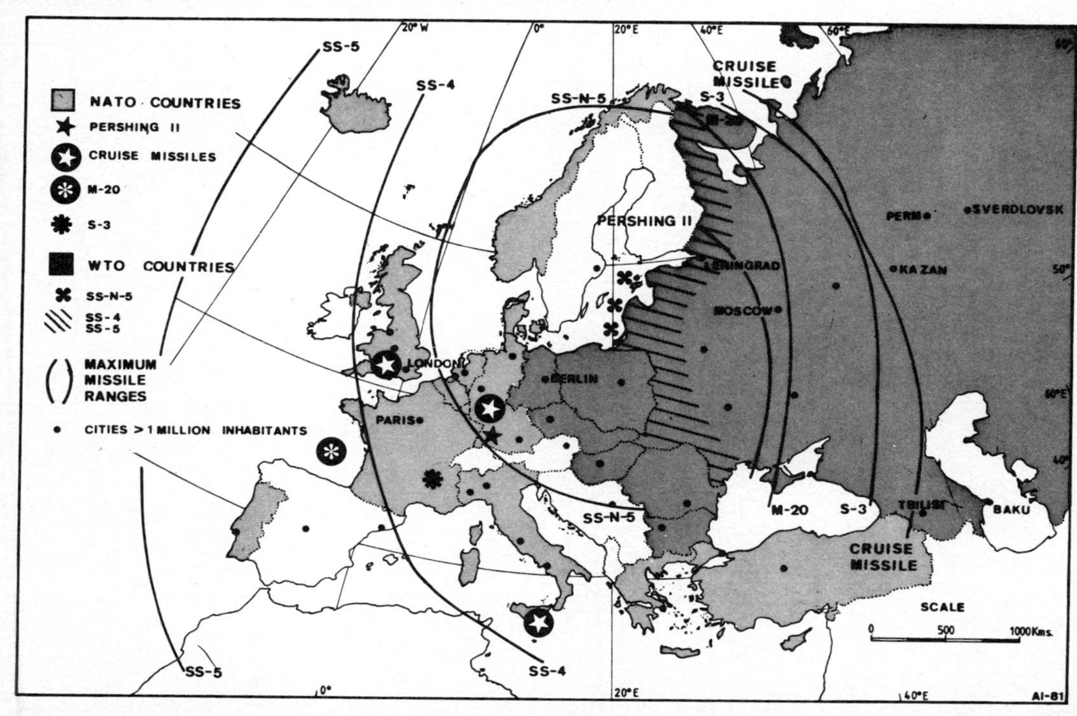

Baltic, four with the Northern Fleet and three in the Pacific. Each carries three SS-N-5 Serb SLBMs, with a range of 1 200 km and a megaton-yield warhead.

The NATO arsenal

The United States. The United States plans to resume its forward deployment of LRTN missiles in Western Europe with the introduction of Pershing IIs and ground-launched cruise missiles (GLCMs).

The Pershing I was operational in 1962 with a range of 650 km, later extended to 740 km for the Pershing IA. Having a CEP (circular error probability) of somewhat less than 400 m and a W.50 warhead with a variable yield of 60–400 kt, extensive collateral damage was unavoidable, while some hardened targets, such as C^3 bunkers, could not be destroyed. Development of the Pershing II began in April 1974, centred on improving accuracy through terminal guidance. In 1978 the range requirement was extended to 1 800 km, and the full-scale development contract was signed with the Martin Marietta Company in February 1979.

The accuracy achieved by the RADAG (radar area guidance) terminal guidance system is the best of any ballistic missile. In the fifth test shot, in May 1978, the warhead impacted within 25 m of the designated target. In the terminal phase, radar returns are compared with a reference image stored in the guidance computer, and position errors are then corrected. The reference image is based on the surroundings of the target, so that the missile is not deceived by camouflage or by a target buried underground [7].

Several aspects of the Pershing II—among them, the warhead—had not been finally determined by the end of 1981. Selectable yields down to 1 kt have been mentioned. Its range, accuracy and short response and flight times make it an extremely versatile and potent weapon. The pre-launch survivability is enhanced through readiness, part of the force being on quick-reaction alert; a Pershing II Firing Platoon can count down and fire three missiles simultaneously[5]; and the Automatic Reference System does not require the Pershing to be launched from presurveyed sites [8].

The USA at present operates 108 Pershing IA missile launchers in its 56th Field Artillery Brigade, the headquarters being in the southwestern part of FR Germany (three battalions of 36 launchers each in Neckarsulm, Schwäbisch-Gmünd and Neu-Ulm). The plan is to replace them with an equal number of Pershing II launchers, beginning at the end of 1983. For the Pershing IA, reload missiles exist; the same will probably be the case for the Pershing II. Plans are for the deployment to be completed

[5] The Pershing missiles are organized in battalions of 36 missile launchers, with four firing batteries (nine launchers each), each battery in turn consisting of three firing platoons (three launchers each).

by 1985; however, the first flight-test of the Pershing II extended range is not scheduled until July 1982.[6]

In addition, the West German *Luftwaffe* operates two Pershing IA wings—FKG 1 at Landsberg and FKG 2 at Geilenkirchen—each with 36 launchers. These may be replaced by the original version of the Pershing II, that is, with RADAG but without extended range, and will therefore remain in the MRTNF category.

The largest increase in the number of US nuclear warheads since MIRVing may result from the massive cruise missile programme. The air-launched (AGM-86B) and sea-launched (BGM-109 Tomahawk) versions are based on the same propulsion and guidance techniques, sharing the characteristics of long range, mobility, penetrability (flying 30 m above the ground and with a radar cross-section of 0.05 m^2, or one-thousandth that of a B-52) [10], and high accuracy (in the region of 50 m).[7] The TERCOM (terrain contour matching) guidance system—basically radar-updated inertial—may be supplemented with terminal guidance. One such system, the DSMAC (digital scene-matching area correlator) is currently being developed for the conventional land-attack version of the SLCM (the TLAM-C).

The GLCMs will be organized in so-called flights of four TELs (transporter, erector, launcher), each with four missiles. They will be housed in facilities hardened against conventional attack. Under a nuclear threat, they rely on mobility and dispersion for pre-launch survivability. The planned inventory of 464 GLCMs will be distributed as shown in table 8.2.

The GLCM programme is small compared with the deployment plans for air-launched and sea-launched cruise missiles (ALCMs and SLCMs). A total of 3 780 ALCMs are now on order [12]. The US Navy plans to procure a total of 3 994 SLCMs for land-attack and anti-ship missions, partly with conventional and partly with nuclear warheads [13]. In his October 1981 statement on strategic policy, President Reagan announced the deployment of several hundreds of nuclear-armed SLCMs on attack submarines, beginning in 1984. The Vertical Launch System will make every major US naval vessel a potential strategic nuclear factor: the eight remaining Polaris submarines may take up to 80 Tomahawks; if the battleships of the Iowa-class (BB-61) are refurbished, they may initially take 32 and later up to 320; the CG-47 Ticonderoga-class (Aegis) guided missile cruisers, 122; the DD-963 Spruance-class destroyers, 61; and the SSN-688 Los Angeles-class hunter-killer submarines, 12 each.

[6] The testing of an operational, mobile Pershing II will reportedly not take place until mid-1983. However, the RADAG guidance system is believed to have been adequately tested on the short-range version [9].

[7] Doubts have been expressed about whether the performance goals of the Pershing II and GLCM will be fully achieved [11].

Table 8.2. Planned deployment of ground-launched cruise missiles in Europe

Country	Base	Number	Year of initial operational capability
United Kingdom	Greenham Common, Berkshire	96	1984
	Molesworth, Cambridgeshire	64	1988
FR Germany	Probably Ramstein, Hahn, Spangdahlem, Brüggen and Laarbruch	96	1984
Italy	Vicenzo Magliocco, Comiso, Sicily	112	1984
Netherlands	..	48	..
Belgium	..	48	..

The United Kingdom. The four British Resolution-class SSBNs (ballistic missile submarines) were phased in from 1967 to 1969. They are equipped with 16 US-delivered Polaris A-3 missiles, each with three British-built MRV (multiple re-entry vehicles, not independently targetable) warheads with a yield of 200 kt. In order to ensure the penetration of ABM (anti-ballistic missile) defences, the United Kingdom has developed the 'Chevaline' system for its Polaris missiles, probably with six manoeuvrable warheads of 40–50 kt each. In the meantime, the Soviet Galosh ABM system around Moscow has had its missile launchers reduced from 64 to 32 [14].

The British government has decided to replace the Polaris-equipped SSBNs, starting in the early 1990s. The precise scope of this programme has yet to be decided, both as regards the number of SSBNs (four or five), the number of SLBMs per submarine (16 to 24), and the type of missile: Trident I (C-4) with eight MIRVs or Trident II (D-5) with up to 14 MIRVs (14 is the SALT II limit; technically, more are feasible). This means that the number of sea-based warheads could range from 512 to more than 2 040 warheads. A likely number is 640 (four submarines with 16 Trident IIs carrying 10 warheads each)—a tenfold increase in the number of independently targetable nuclear warheads.[8] Economic constraints, however, attach a measure of uncertainty to the whole programme.

France. Apart from the USA and the USSR, France is the only country to operate a full strategic triad of land-, air- and submarine-based nuclear weapons.

The weakest leg is the small force of silo-based missiles at the Plateau d'Albion near Avignon. One squadron has nine SSBS (*Sol-Sol Balistique*

[8] UK Defence Secretary Nott has hinted that the US decision to go ahead with the D-5 missile means that the UK almost certainly will adopt it [15].

Stratégique) S-3s, with a range of somewhat more than 3 000 km and a 1-Mt warhead. The other squadron is in the process of converting from the older S-2, with a 150-kt warhead, to the S-3.

France increasingly relies on SSBNs. The FOST (*Force Océanique Stratégique*) now operates five SSBNs, each with 16 MSBS (*Mer-Sol Balistique Stratégique*) M-20 SLBMs, with a range of some 3 000 km and a 1-Mt warhead. A sixth submarine is due to enter service in 1985 with the MRVed M-4 missile (4 000 km range, 6 to 7 150-kt warheads). The entire fleet will convert to the M-4 by 1989.[9] A seventh SSBN of a new class and with M-5 missiles—probably MIRVed—will join the fleet in 1994 at the earliest [16].

A mobile, land-based ballistic missile—the S-X—is scheduled to replace the Mirage IVA aircraft in the 1990s.

Aircraft

Aircraft have several disadvantages when compared to ballistic missiles in an LRTN role, the most significant being longer flight times, pre-launch and in-flight vulnerability. The vulnerability problem is severe for both sides: the WTO has a dense and overlapping surface-to-air missile (SAM) and interceptor network, whereas NATO, while improving SAMs, concentrates on AWACS (E-3A and Nimrod AEW) and advanced interceptors/air superiority fighters (F-15/Tornado ADV). The air defence environment is especially dense in the Central European region, which may force strike aircraft to operate to an increasing extent on the flanks [17].

On the other hand, aircraft do have certain advantages over missiles: they can carry large and diversified weapon loads, can attack several targets on the same mission—including mobile 'targets of opportunity'—can observe the results of their own strikes and those of others, can achieve high accuracy, and can be recalled after the take-off. In order to increase the range, to allow for the use of air-to-surface missiles (ASMs) (which are heavier and larger than free-fall bombs), and so on, actual weapon loads are, however, likely to be considerably lower than the potential maximum.

The comparisons presented in tables 8.3–8.7 are based on the following criteria:

(*a*) Because of the difficulties in determining the number of aircraft actually nuclear-configured, all aircraft of types that are nuclear-capable have been included.

[9] The first French SSBN, *Le Redoutable* (laid down in 1964, operational in 1971), which will come to the end of its operational life around the mid-1990s, may not convert to the M-4.

(*b*) Ranges for specific mission profiles are classified, and most sources do not specify the conditions for the ranges given. SIPRI estimates are based on the high–low–high profiles (low-level final approach to target). The possibility of in-flight refuelling has not been taken into account, although this could increase the range considerably. Most NATO aircraft are equipped for refuelling; NATO also has more aerial tankers than the WTO, and seems more proficient in using them. Combat radii are given, although the aircraft could return to other airfields than those they started from, or even be sent on a one-way mission. (For further details, see the notes to table 8.4.)

Table 8.3. Official estimates of long-range theatre nuclear aircraft in Europe

Estimates	Western aircraft		Soviet aircraft	
US figures[a]	FB-111	63[b]	Tu-22M	45
	F-111	154	Tu-16/Tu-22	350
	F-4	265	Fencer/Fitter/Flogger	2 700
	A-6/A-7	68		
Total		**560**		**3 095**
Soviet figures[a]	FB-111	65[b]	Tu-16/Tu-22/Tu-22M	461
	F-111	172		
	F-4	246		
	A-6/A-7	240[c]		
	Vulcan B.2	55		
	Mirage	46		
Total		**824**		**461**

[a] Sources are the same as those given in note *a* to table 8.1.
[b] Based in the USA, but intended for use in Europe.
[c] Presumably aircraft on US carriers in the Mediterranean Sea and the Atlantic Ocean.

(*c*) The tables list the total number of aircraft, including reserves and aircraft in training units. This leads to inflated numbers as regards aircraft actually available for any single mission, but gives comparable numbers for the WTO and NATO and is the principle used in SALT [18].

The aircraft are divided into two categories: (*a*) *primary* LRTN aircraft, with combat radii well over 1 000 km, and with a low-level, all-weather capability to ensure penetration (the Panavia Tornado (MRCA) has, for instance, been put in this category because of its excellent low-altitude, all-weather capability, although its range would more properly place it in category *b*); and (*b*) *marginal* LRTN aircraft, with combat radii of about 800–1 200 km, and a limited low-level, all-weather capability. The F-16 Fighting Falcon has been placed in this category, *inter alia*, for lack of an all-weather capability.[10]

[10] From the mid-1980s, F-16s will be equipped for day/night all-weather operations, with navigation/target location from satellite or aircraft. In addition, there are plans for an extended-range version, the F-16E (XL), with all-weather capabilities [19].

Table 8.4. Primary long-range theatre nuclear aircraft

Country	Designation	Year first deployed[a]	Combat radius (km)[b]	Inventory[c] Total	Inventory[c] European[d]	Programme status
USSR	Tu-22M Backfire	1974	3 000	75	60	Production rate: up to 30 per year, half of them assigned to naval aviation
	Tu-16 Badger	1955	2 000	300	225	
	Tu-22 Blinder	1962	1 200	130	100	
	Su-24 Fencer	1974	1 700	500	375	Production rate: approx. 60 per year
USA	FB-111A	1969	1 800	63	0	
	F-111A/D/E/F	1967	2 000	300	156	
UK	Vulcan B.2	1957	2 700	55	55	Being replaced by Tornado
	Tornado GR.1 (IDS)[e]	1982	1 400	0	0	220 programmed (incl. 68 dual-control trainers); last 20 may be converted to F.2 (ADV)
France	Mirage IVA	1964	1 600	35	35	More than 15 will continue in service after 1985
	Mirage 2000N	1986	1 400	0	0	Up to 200 may be acquired
Rest of NATO	Tornado IDS	1982	1 400	0	0	FR Germany plans 212 (incl. 47 dual-control trainers); Italy plans 100 (incl. 12 trainers)

[a] Date for deployment of first version in country of origin.

[b] Ranges assume a high–low–high mission profile (with low-level, high-speed final approach to the target), maximum external and internal fuel, but no in-flight refuelling, and that the payload includes external nuclear ASMs where applicable. The ranges of the ASMs are, however, not added to that of the aircraft.

The given ranges are *maximum* combat radii, which might be reduced by the need for evasive action, fuel reserves (for landing and loiter), external ECM equipment (which reduces fuel load and increases aerodynamic drag), more demanding mission profiles to increase penetration and survivability, etc.

[c] Numbers given are total, i.e. including all aircraft of types that are considered dual-capable, covering aircraft in the maintenance cycle.

Trainers are excluded (save dual-control versions of aircraft that are two-seaters in their basic version), and reconnaissance aircraft (unless they are basic versions equipped with pods).

Actual numbers of nuclear-configured, mission-ready aircraft are substantially lower.

[d] Aircraft based in Europe or within striking range of targets in Europe without refuelling. For the USSR, this is estimated at three-quarters of the total.

[e] Tornado GR.1 is the British designation of the Panavia Tornado IDS (interdiction/strike version). The United Kingdom also plans to acquire 165 of the air defence variant (ADV) of the Tornado, with the official designation Tornado F.2.

Table 8.5. Marginal long-range theatre nuclear aircraft

Country	Designation	First deployed[a]	Combat radius (km)[b]	Inventory[c] Total	Inventory[c] European[d]	Programme status
USSR	MiG-23/27 Flogger	1971	900	2 000[e]	1 500	Production continues at 500 per year (incl. exports)
Rest of WTO	MiG-23 Flogger	1971	900	200	200	
USA	F-16 Fighting Falcon	1979	1 300	300	0	Total programme: 1 388 (incl. 204 F-16B trainers); more than 200 will be deployed in Europe
	F-4 Phantom II	1961	1 100	1 400	250	Being phased out of active duty and transferred to the Reserve Force
	A-7 Corsair II	1966	1 200	370	0	Reserve Force
UK	Buccaneer S.2	1962	1 400	60	60	Excl. 20 in maritime strike role (cf. table 8.6)
	Jaguar GR.1	1973	1 200	140	140	Excl. 30 Jaguar T.2 trainers
	Harrier GR.5 (AV-8B)	1986	900	0	0	Total programme: 60
France	Jaguar A	1973	1 200	160	160	Total number procured (losses unknown); excl. 40 Jaguar E trainers
	Mirage IIIE	1961	1 000	135	135	Being phased out; excl. 14 Mirage IIIBE trainers
Rest of NATO[f]	F-16 Fighting Falcon	1979	1 300	64	64	Excl. 8 F-16B trainers; total programme: 194 F-16A (of which 4 have been lost) and 46 F-16B (2 lost)
	A-7H/P Corsair II	1966	1 100	63	63	Excl. 6 TA-7H trainers; 11A-7P on order for Portugal
	F-4E Phantom II	1961	1 100	134	134	Incl. 10 West German F-4E in USA for training; excl. 96 RF-4E and 168 F-4F
	F/CF-104G Starfighter	1958	1 000	555	525	Incl. 30 West German F-104G in USA for training; excl. 145 TF-104G and RF-104F; being phased out
	F-104S (Aeritalia)	1969	1 000	196	196	

[a] Date for deployment of first version.

[b] Ranges assume a high–low–high mission profile (with low-level, high-speed final approach to the target), maximum external and internal fuel, but no in-flight refuelling, and that the payload includes external nuclear ASMs where applicable. The ranges of the ASMs are, however, not added to that of the aircraft.

The given ranges are *maximum* combat radii, which might be reduced by the need for evasive action, fuel reserves (for landing and loitering), external ECM equipment (which reduces fuel load and increases aerodynamic drag), more demanding mission profiles to increase penetration and survivability, etc.

[c] Numbers given are total, i.e., including all aircraft of types that are considered dual-capable, covering aircraft in the maintenance cycle.

Trainers are excluded (save dual-control versions of aircraft that are two-seaters in their basic version), and reconnaissance aircraft (unless they are basic versions equipped with pods).

Actual numbers of nuclear-configured, mission-ready aircraft are substantially lower.

[d] Aircraft based in Europe or within striking range of targets in Europe without refuelling. For the USSR, this is estimated at three-quarters of the total.

[e] Including 600 MiG-27 Flogger Ds, but excluding some 1 000 MiG-23s in the air defence force (*PVO-Strany*), which are not considered to be nuclear-capable.

[f] Excludes Canadian, Danish and Norwegian aircraft, which are unlikely to be converted to nuclear roles.

Table 8.6. Naval long-range theatre nuclear aircraft

Country	Designation	Year first deployed[a]	Combat radius (km)[b]	Inventory[c] Total	Inventory[c] European[d]	Programme status
USSR	TU-22M Backfire	1974	3 000	75	60	Naval aviation has received half the number of Backfires
	TU-16 Badger	1955	2 000	250	190	
	TU-22 Blinder	1962	1 200	50	35	
USA	A-6 Intruder	1963	1 500	250	20	
	F-18 Hornet	1982	1 100	0	0	Total programme: 1 377 (including TR-18 trainers)
	A-7 Corsair II	1966	1 200	360	48	Being replaced by F-18
	F-4 Phantom II	1961	1 100	200	0	Excl. 200 non-nuclear Marine Corps F-4s; being replaced by F-18s
	AV-8B Harrier II	1985	900	0	0	Total programme: 322
UK	Buccaneer S.2	1962	1 400	20	20	Approx. number dedicated to CINCLANT; will continue for some time after the rest of the Buccaneers are replaced by Tornado IDS
France	Super Etendard	1979	700	60	60	The 300-km range of the ASMP will give it marginal long-range theatre nuclear capability; total programme: 71
FR Germany	Tornado IDS	1982	1 400	0	0	Total programme: 112 (including 10 dual-control trainers)
	F-104G Starfighter	1958	1 000	95	95	Excl. 10 TF-104G and 24 RF-104G; being replaced by Tornado

[a] Date for deployment of first version in country of origin.

[b] Ranges assume a high–low–high mission profile (with low-level, high-speed final approach to the target), maximum external and internal fuel, but no in-flight refuelling, and that the payload includes external nuclear ASMs where applicable. The ranges of the ASMs are, however, not added to that of the aircraft.

The given ranges are *maximum* combat radii, which might be reduced by the need for evasive action, fuel reserves (for landing and loitering), external ECM equipment (which reduces fuel load and increases aerodynamic drag), more demanding mission profiles to increase penetration and survivability, etc.

[c] Numbers given are total, i.e., including all aircraft of types that are considered dual-capable, covering aircraft in the maintenance cycle.

Trainers are excluded (save dual-control versions of aircraft that are two-seaters in their basic version), and reconnaissance aircraft (unless they are basic versions equipped with pods).

Actual numbers of nuclear-configured, mission-ready aircraft are substantially lower.

[d] Aircraft based in Europe or within striking range of targets in Europe without refuelling. For the USSR, this is estimated at three-quarters of the total. For the USA, aircraft on board 2 carriers (2 Carrier Air Wings) have been included.

Table 8.7. Air-to-surface missiles[a]

Country	Designation	Year first deployed	Range (km) (high-level launch)	Warhead	Speed (Mach)	Inventory	Notes, programme status
USSR	AS-2 Kipper	1961	210	1 × kt-range/HE	1.2	n.a.	1 × Tu-16
	AS-4 Kitchen	1962	720	1 × kt-range	2.5	135	1 × Tu-22 2 × Tu-22M
	AS-6 Kingfish	1977	700	1 × 200-kt	3	65	1 × Tu-16 2 × Tu-22M
USA	AGM-69A SRAM	1972	160	1 × 170-kt	3	378[b]	6 × FB-111A
France	ASMP	1985	300	1 × 150-kt	3	0	1 × Mirage IVA (1985) 1 × Mirage 2000N (1986) 1 × Super Etendard (1987) Total programme: 100

[a] Nuclear-capable ASMs being used on LRTN aircraft.

[b] Maximum force loading for the FB-111A. Total SRAM inventory 1 250.

The use of ASMs may significantly increase the penetration capability of aircraft systems. This is especially important for old, vulnerable aircraft such as the Soviet Tu-16 and Tu-22. In addition, ASMs increase the range of the systems; thus, the French Super Etendard carrier-based naval aircraft achieves a range that puts it in the marginal LRTN category when using the 300-km range ASMP (*Air-Sol Moyenne Portée*) missile. France will equip its Mirage IVA, Mirage 2000 N and Super Etendard aircraft with this missile. The USA operates the short-range attack missile (SRAM) on its FB-111As, and may develop a relatively short-range cruise missile to fit smaller aircraft. The technical feasibility of equipping the Tornado with cruise missiles has been explored [20].

The WTO arsenal

Primary LRTN aircraft. The Soviet Long-Range Aviation (LRA) operates several hundred medium-range bombers, the most numerous of which is still the Tu-16 Badger. In terms of capabilities it may be compared to the US B-47, which was phased out in the mid-1960s. Badgers can be expected to remain in service until about the end of this decade, though increasingly converted to reconnaissance, electronic countermeasures (ECM) and tanker configurations. Some 300 Badgers are currently in service in the bomber role, with another 100 or so in support roles. In addition to the Badgers, the LRA operates some 130 Tu-22 Blinders, also approaching obsolescence.

The Soviet LRTN bomber force is increasingly based on the Tu-22M Backfire. The combat radius of the Backfire bomber is sufficient to reach any European target from bases in the Soviet Union, but it may have to depend on ASMs to deliver its weapons to the target. About 75 are operational with the LRA forces.

One of the most important developments in the Soviet aircraft arsenal is the steady increase in the number of Su-24 Fencers, which have capabilities that place them somewhere between the Tornado and the F-111. Some 500 Fencers are operational so far, and production continues at a rate of 60 or more per year. None is based outside the Soviet Union.

Marginal LRTN aircraft. The MiG-23/27 Flogger is becoming the standard Frontal Aviation fighter. The number of Floggers in service is increasing very rapidly, more than 500 being produced every year (including those for export). Today, some 2 000 Floggers are operational, including 600 of the MiG-27 Flogger D ground-attack version. However, when considering the LRTN potential for the large Flogger force, it should be borne in mind that it has a limited all-weather capability, and that even the MiG-27 versions are range-restricted and primarily intended for close support of ground forces. Other WTO countries are also phasing in the MiG-23, mostly the MiG-23BM Flogger F, which is a somewhat simplified export version of the MiG-27 Flogger D. So far, these countries have some 200 of them.

Western estimates of the LRTNF balance often include other dual-capable aircraft as well, such as the MiG-21 Fishbed and the various Fitter versions (Su-7/17/20). Even the most long-ranged of these, the swing-wing Su-17/20, has an operational combat radius that excludes it from LRTN calculations.

Finally, some allowance for Soviet Naval Aviation forces seems justified. While primarily intended for anti-ship roles, their potential for land attack is obvious. Some 75 Tu-22Ms (half of the Backfire force), 250 Tu-16s, and 50 Tu-22s have been assigned to Naval Aviation.

The NATO arsenal

The United States. The US Air Force in Europe (USAFE) has 500 combat aircraft at its disposal. The most potent LRTNF component is the two F-111 wings based in the UK, with 66 F-111Es and 90 F-111Fs. Other dual-capable aircraft based in Europe, in the marginal LRTNF category, include some 250 F-4 Phantoms—with the F-16 being phased in.

In times of crisis or war, these forces can be greatly expanded by transfer of CONUS-based aircraft to 43 Collocated Operating Bases and 14 Main Operating Bases in Europe, amounting to a total of 960 combat aircraft, and another 592 if bases are available [21].

The total US inventory figures include aircraft already in Europe, those based in Asia, as well as Air Force Reserves and the Air National Guard. The transfer of aircraft from Asia is a remote possibility, while the general significance of Reserves and National Guard forces is increasing.

Regarding the number of CONUS-based F-111s that are available for European contingencies, it is assumed that at least 144 F-111A/Ds still exist. Ninety-six of them are declared to be in active service. Of the strategic FB-111A version, 63 aircraft exist. These are not SALT-accountable, and are clearly intended for missions in Europe. Both sides include them in their LRTNF estimates.

The Phantom is being transferred to the reserves while the F-16 Fighting Falcon is being phased in; of a total production order of 1 182 (and another 206 F-16B dual-seat trainers), some 300 have been delivered. The 370 A-7D Corsair IIs are all in the Reserve Force.

In US naval aviation, the F-18 Hornet will become the standard fighter aircraft in the years ahead (supplemented by the F-14A Tomcat in the air defence role). It replaces the A-7E Corsair II and the F-4 Phantom II. The all-weather-capable A-6 Intruder will be maintained for long-range strike missions. The AV-8B Harrier II is scheduled to become operational in 1985.

The United Kingdom. The 55 Vulcan B.2 will be replaced at a rapid rate by the Tornado GR.1, which is entering service from 1981–82. The same applies to the Buccaneer S.2. By 1986, all Vulcans and Buccaneers should have been replaced by the Tornado interdiction/strike version. Jaguar GR.1 aircraft are also nuclear-capable, entering the marginal LRTN category.

France. The longest-serving leg of the French strategic nuclear triad is the Mirage IVA, with a total of 62 delivered by 1968. Thirty-five remain in the bomber role (including two in reserve). They are supported by 11 KC-135F aerial tankers. More than 15 Mirage IVAs will be kept beyond 1985, re-equipped with the ASMP stand-off missile in exchange for its present AN-22 free-fall 70-kt bomb.

The Tactical Air Force operates Jaguar A and Mirage IIIE aircraft, the nuclear-dedicated version carrying a single AN-52 free-fall 25-kt bomb. These are marginal LRTN aircraft. The strike version of the next-generation fighter, the Mirage 2000N, armed with ASMPs, will replace the Mirage IIIE from 1986–87, and is considered to be a primary LRTN aircraft.

Other NATO countries.[11] The most important LRTN aircraft in the other NATO countries is the F-104G Starfighter, especially for FR Germany. The Starfighter, which was introduced in the early 1960s, is now being replaced by the Tornado (in FR Germany and Italy), and the F-16 (in Belgium and the Netherlands), but will remain in service for some time in Greece and Turkey.

Canadian, Danish and Norwegian aircraft are not taken into account, as they are unlikely to be converted to nuclear configuration.

Force ratios

As far as weapon systems within striking range of Europe are concerned, the force ratios are roughly as follows.

In the missile sector, the Soviet Union has a predominance in the number of launchers, of the order of 2.5:1 (if US figures are accepted) or 2:1 (using Soviet numbers). The disparity appears in the figures for the remaining SS-4s and SS-5s.

For primary LRTN aircraft (including CONUS-based FB-111As), the ratio is 2.5:1, for a WTO advantage. The inclusion of naval aircraft does not change that ratio significantly. Towards the end of the decade, this numerical advantage is likely to be somewhat reduced—even with the continued production of Backfires and Fencers at present rates—as the

[11] Spain will probably join NATO, which will add the following dual-capable aircraft: 19 Mirage III-EEs (plus six trainers) and 37 F-4C Phantom IIs. On the other hand, Greece may withdraw its participation in NATO's nuclear posture.

Tornado enters service and the Tu-16/22s reach the end of their serviceable lifetime.

There are wide disparities in the official figures for the aircraft sector. Apart from the public relations debate over numbers that came to a head before the opening of the Geneva negotiations and other tactical considerations which enter into the calculations, the disparities reflect a variety of difficulties in counting LRTN aircraft. There is bound to be a certain arbitrary element in any estimate.

If warheads are counted, the Soviet Union has a much greater predominance in the missile sector, as long as British and French forces are not MIRVed. For aircraft, weapon loads are too flexible and uncertain for overall estimates to be meaningful.

IV. Theatre nuclear doctrines

Soviet LRTNFs, in particular the SS-20, have created a sense of inferiority and insecurity in Western Europe and may be used for purposes of intimidation and blackmail. In times of war, Soviet doctrine emphasizes initiative, surprise, deep strikes and massive use, which can now be executed with greater precision than before. It is small comfort that this has been the Soviet doctrine for 20 years, that Western Europe has lived under the shadow of Soviet LRTN missiles ever since the end of the 1950s or beginning of the 1960s, and that to some extent the threat was only 'rediscovered' in 1977 with the deployment of the SS-20s and the general deterioration of East–West relations.

An historical perspective is indispensable for any assessment of the political and military functions of these systems. While gross Soviet inferiority in intercontinental systems and the 'holding Europe hostage'-factor belong to the past, Soviet LRTNFs are still opposing US forward-based systems. The FBSs have been on the decline for some time, but may again increase considerably through the deployment of Pershings and the wide dispersal of cruise missiles. In addition, French, British and Chinese forces are growing. The Soviet Union has a number of regional security concerns, each with its own specific military aspects.

In all likelihood, a domestic momentum of an industrial, a bureaucratic and a military nature has also influenced the genesis and scale of present programmes, giving the Soviet Union a current numerical lead of more than 2:1 in LRTN systems within striking range of Europe. In the West, where much higher ratios are mentioned, this has produced considerable concern and anxiety—all the more so since, in political life, there is often no strict relationship between cause and effect. Nor are history and the international context always properly taken into account.

The High Level Group (HLG), established by NATO in October 1977 to study the need for new LRTNFs, agreed that NATO's modernization decision should reflect an evolutionary change in the alliance's posture, with no change in nuclear strategy and no change in the overall number of nuclear weapons in the European theatre [22]. The deployment would not be required to match the number of SS-20s and other Soviet LRTN systems, but should be sufficient to keep open the option of striking a substantial number of Soviet targets from Western Europe, the military/technical judgement being that existing forces were no longer adequate for that purpose. Hence, the military rationale for new missiles was based on the need to enhance the coupling between theatre forces and US intercontinental systems, reinforcing the US nuclear umbrella over Western Europe. It was undoubtedly a response to Soviet modernization as well, but politically more than militarily, providing bargaining leverage for negotiations with the Soviet Union [23]. In public discussions, the need for new LRTNFs has largely been ascribed to the Soviet build-up of SS-20s and Backfire bombers. This is superficial: rather, the SS-20 has been singled out for particular attention to justify a perceived need to modernize which, sooner or later, would have arisen in any case [24].

Making the SS-20 the big public argument for new cruise and Pershing missiles is a double-edged sword. To emphasize the SS-20 as a source of insecurity and potential blackmail to the extent that some political leaders have done amounts to declaring oneself open to pressure already in advance. In this connection, it may be worthwhile recalling that self-fulfilling prophecies are not uncommon in politics [25]. On the other hand, if West European politicians declare that they are not susceptible to nuclear blackmail, it is not clear what the Soviet Union could blackmail Western Europe into doing. The whole blackmail theory needs a serious examination [26].

Soviet military doctrine starts from the premise that if another war occurs in Europe, it should be fought as far towards the West as possible. Nuclear, chemical and conventional weapons are highly integrated, and Soviet strategic literature does not emphasize the selective use of nuclear weapons and the limitation of collateral damage as NATO declaratory strategy does. Generally, the logic of Soviet military doctrine seems coherent. NATO doctrine, however, is based on premises which have been heavily criticized for lack of consistency and credibility.

The military-strategic rationale for new NATO missiles

The official military-strategic justification for the deployment of cruise and Pershing missiles in Western Europe hinges on the 'coupling' argument and the maintenance of the US nuclear umbrella over Western

Europe. In the following sections, an attempt is made to examine these arguments by first elaborating, in some detail, two alternative rationales for the new missiles along those lines—without pretending that Western defence officials would subscribe to all of it—and then assessing the validity of the assumptions.

If cruise and Pershing missiles are forward-based in Western Europe, pressure will arise to fire them before they are captured or destroyed, or in retaliation to a Soviet attack with similar weapons. Both cruise and Pershing missiles would reach targets on Soviet territory. With a range of 1 800 km, the Pershing missiles would reach almost as far as Moscow from their deployment positions in the Schwäbisch-Gmünd–Neu Ulm–Neckarsulm area.[12] There would, in other words, be a US nuclear attack on the Soviet Union, the likely response to which is Soviet retaliation against US territory, that is, the escalation of warfare to the strategic level. This is consistent with traditional West German and other NATO declaratory policy: for the European countries, nuclear weapons are primarily political weapons—their only rational function being that of dissuasion by deterrence—and a credible threat of rapid escalation to the strategic level would be the most effective deterrent. Cruise and Pershing missiles serve precisely that function because they will couple the theatre nuclear forces with the intercontinental systems of the United States. Therefore, the US nuclear umbrella over Western Europe—questioned ever since the advent of Soviet intercontinental missiles, and increasingly so as the Soviet Union achieved rough nuclear parity—would be re-inforced or re-established.

Land-based missiles are, furthermore, more effective in the coupling role than sea-based missiles would be. Submarine-based weapons could be held in relatively invulnerable positions for long periods of time, so the pressure to use them at an early stage might be less and the escalatory effect therefore more uncertain. Moreover, land-based missiles are more visible than sea-based ones and therefore also more credible couplers of US and European destinies in the public eye. This psychological-political argument loomed high in the justification for land-basing before the so-called dual-track decision was made on 12 December 1979.

The basing areas in FR Germany are such as to maximize the coupling effect. Like the Pershing II, the cruise missiles will move around in the western parts of FR Germany; if the scenario is that of a European war between the two alliances, it is therefore reasonable to assume that nuclear weapons will already have been used before Soviet forces eventually reach the deployment areas of cruise and Pershing missiles, that is,

[12] Awaiting flight-testing of the extended-range version, the precise range of the missile is not known. A further developed version of the Pershing with a range of about 4 000 km is in "technology development", but so far under limited funding.

before they have to be fired.[13] The first use of nuclear weapons—a very hard decision to make—is therefore likely already to have occurred, and the further use of nuclear arms is usually assumed to be less difficult to authorize. Since the launching of cruise and Pershing missiles is not likely to be a question of first use, the firing of them becomes more thinkable and, consequently, more likely actually to happen. Seen from the USSR, the threat of retaliation against Soviet territory therefore becomes more credible and the deterrence effect all the more formidable —or so the reasoning goes.

The deterrence threat of retaliation against Soviet territory may also be enhanced in another way. It is sometimes hypothesized that it may work in this manner: if highly accurate nuclear missiles are launched against the USSR from Western Europe, the Soviet Union would retaliate against Western Europe and not against US territory, for fear that its less accurate missiles would lead to an all-out strategic war if launched against the United States. And the more likely it becomes that Soviet retaliation will be directed against Western Europe, the higher the probability is that the USA will actually use the new cruise and Pershing missiles against targets on Soviet soil. Thus, Soviet territory would not be a sanctuary in a European war. The Soviet leaders would know from where the attack is launched—and by implication, also where to retaliate. This proposition is clearly incompatible, however, with the rationale outlined above.

A key factor in this line of reasoning is high accuracy. The CEP for Pershing and cruise missiles is only about 40–50 metres, so even with a low-yield nuclear warhead, the Pershing will be a very effective counterforce weapon, and the collateral damage may be relatively low. The Soviet Union, which is unable to retaliate with similar high-accuracy, low-yield weapons against US territory, may therefore respond by turning its less accurate weapons against Western Europe. In this case, Western Europe would be the hostage and eventually the victim of a US nuclear attack on the European part of the USSR—a situation similar to that which existed before the advent of Soviet intercontinental forces, when Soviet LRTNFs were deployed to compensate for the 'missile gap' (which was real enough, but in the US favour). The other edge of this deterrent sword is therefore the Europeanization of nuclear war on US terms.

This is not the first time the United States has tried to capitalize on its lead in missile accuracy to bolster the European belief in the nuclear umbrella. When former Secretary of Defense Schlesinger presented his nuclear weapon targeting and employment policy in 1974—spelled out in

[13] In most of the war game scenarios played out by NATO, 8-in howitzers and 155-mm batteries are fired first [27].

National Security Decision Memorandum 242—the selective use of strategic weapons against Soviet territory was one of the new features. The selective options were justified by reference to the need to reinforce the umbrella over Europe [28]. Schlesinger preferred Minuteman ICBMs for execution of such options; SLBMs might have been as accurate (*inter alia* because of shorter flight distance to target) but not as reliable. Anyhow, accuracy was a key factor in the strategy of selective options, and since the USSR could retaliate only by means of less accurate missiles with higher-yield warheads, it was assumed that the selective options strategy was credible. An attack on Western Europe could, in other words, lead to the use of US strategic weapons in a selective mode. In conclusion, the strategy would create a stronger link between TNFs in Europe and US intercontinental systems, thereby enhancing deterrence.

The new feature in relation to the cruise and Pershing missiles is, therefore, not the effort to capitalize on the lead in missile accuracy, but the forward basing of the missiles in FR Germany, Italy, the United Kingdom, and eventually other European countries. Land-basing is more visible than sea-basing also in the sense that the Soviet Union will be in a better position to determine which types of forces are used, namely, LRTNFs from Western Europe or strategic forces from the sea, covered by SALT. This fits the second deterrence mode mentioned above, insofar as it makes it easier to distinguish the beginning of a theatre nuclear war in Europe from the start of a strategic nuclear exchange between the two great powers.

The first line of thought centres on the coupling effect, and largely confirms the NATO military-strategic rationale for deploying new missiles in Europe. The other one emphasizes that a nuclear umbrella may be achieved by means of highly accurate theatre weapons, drawing the western districts of the USSR into the war area while avoiding retaliation against US territory. How credible are they?

The 'nuclear umbrella': the remarkable life of a myth

The role of nuclear weapons in the defence of Western Europe has been problematic ever since the Soviet Union achieved a potent second-strike capability *vis-à-vis* the United States. When deciding to remove France from NATO's military organization, de Gaulle argued that no US President would sacrifice Chicago for Paris. The umbrella was gone. At the time, his judgement was disputed. However, with the advent of strategic parity, more and more politicians and observers drew the conclusion that the nuclear umbrella of the 1950s had become fiction.

For the Soviet leadership, the detonation of a US nuclear weapon on Soviet territory is certainly an act of strategic nuclear warfare. The planned response can hardly depend on the launching point or such physical characteristics of the delivery vehicle as accuracy. Retaliation against US territory must be assumed to follow. Otherwise, the implication is that the Soviet leaders would, in effect, signal to the USA: "Our homeland is divisible: if you shoot at us West of the Urals from Western Europe we will leave your territory intact, but not if you use Poseidon or Minuteman missiles". Not that retaliation against the USA, with the inherent danger of escalating strategic warfare, is necessarily a rational reaction. But the US government can hardly be expected to gamble on the possibility that the Soviet leaders would scrap their planned response and switch to another standard of rationality at the moment of showdown.

In all likelihood, the nuclear umbrella over Western Europe is just as fictitious as Henry Kissinger said it was in his speech in Brussels in 1979 [29]. In fact, all umbrellas seem to be gone, and the one over Western Europe may have been the last one to fold up. No technological fix is likely to revive it. Thus, in response to Schlesinger's selective options, the Soviet Union probably prepared measured counter-attacks against US territory, for instance, detonation of a similar number of nuclear warheads over deserts or sparsely populated areas, to avoid great damage to cities and industrial centres, thereby limiting the escalatory effect of the response. Launching cruise and Pershing missiles from Western Europe makes no basic difference: this would still be a nuclear attack on the Soviet homeland—and a dramatic act of irrationality, since retaliation against the USA is likely to follow. Therefore, neither the one nor the other of the above-mentioned coupling and umbrella assumptions holds water. On the contrary, to use the new missiles against the Soviet Union, and consciously escalate the war to a strategic level, is something the USA would do its utmost to prevent: if there is anything worthy of being called supreme national interest, it must be the desire to keep one's own country outside the area of direct nuclear warfare. To launch an attack from Western Europe on the assumption that the Soviet Union would then conveniently retaliate against Western Europe and the US forces there, rather than against US territory, is therefore wishful thinking, and so obviously so that US decision makers must have clearly understood this for a long time. There is never going to be a mutual understanding between the two great powers on confining a nuclear war to Europe between the Atlantic and the Urals, leaving the United States as a sanctuary.

A European theatre comprising all European countries and excluding the Soviet Union and the USA (which is a European power by invitation) is another matter. Should a war break out between the military alliances in Europe, both the great powers would, of course, do their very best to keep their own territories out of it. Here, the logic is overwhelming; we do not have to read public statements or war manuals to know that this is so. Precisely because the logic is so compelling, there is no need for US–Soviet talks to establish agreement on it either. That commonality of interest works perfectly well by tacit understanding. This is, *nota bene*, not to say that a nuclear war in Europe will actually be so confined—only that the USA and the USSR will try to confine it. No one can know whether they will succeed: technological mishaps, a chaotic battlefield, the breakdown of C^3I facilities and human behaviour under extreme stress defy prediction.

Nor is an ambitious US strategic programme likely to reinstate the nuclear umbrella. Admittedly, the Reagan Administration is aiming at some kind of clearly perceived upper edge in this decade. Depending on the degree and kind of strength that will be achieved, perceptions of US aims, plans and readiness for action are likely to change, and more assertive US behaviour in various parts of the world might be expected. But to reinstate the umbrella—that is, to make it sound realistic that the USA would use, for example, cruise and Pershing missiles against the Soviet Union in defence of Western Europe—much more is needed. The relationship between the two great powers would have to revert to that which existed at the time of the Cuban missile crisis, and this is out of the question.

Suffice it then to add that, apart from first use, the decision to launch nuclear weapons against the major adversary is likely to be the hardest one to make in a nuclear war. For the Soviet Union to try to capture or destroy cruise and Pershing missiles, or to use its LRTNFs against Western Europe in the massive, deep-strike fashion prescribed by Soviet doctrine, would certainly also be a very dramatic act. However, from a great-power point of view, it would be less consequential than escalation to the strategic level. No US superiority is likely to change this in the foreseeable future.

In conclusion, two deterrence effects of new missile deployments nevertheless remain.

1. The missiles will, after all, add some uncertainty regarding Western responses to a WTO attack, and therefore induce additional caution on the Soviet side. Thus, the postulated coupling effect cannot be entirely discarded. This should be all the more emphasized since a number of Soviet statements allude to the view that a nuclear war in Europe will trigger the use of LRTNFs against the Soviet Union, and quickly escalate to the strategic level.

2. More effective coverage of military targets in the smaller East European countries, of Soviet forces, bases and support facilities in particular, also has a deterrent effect. (Apart from GLCMs and Pershing

IIs, this deterrent effect is enhanced by the current modernization of NATO MRTNFs.)

However, with these modifications, the postulated coupling effect basically appears to be a myth.

V. The role of cruise and Pershing missiles in US strategy

Targeting and employment policies

Would there not be pressure on the United States, after all, to fire the missiles before they are captured or destroyed, or in retaliation to a Soviet attack by similar weapons? The answer is yes—the pressure would no doubt be real. So, granted that it is a gargantuan task to prevent a nuclear war in Europe from escalating to the strategic level, why does the United States invest so much—first of all politically, but also economically—to deploy new missiles in Western Europe? What is the point, in terms of *US* interests? What is the *US* military-strategic rationale for wanting the new weapons deployed?

The answer seems simple. Should the scenario be that of a European battle, the missiles might well be fired—but most likely against the smaller East European countries, and preferably against Soviet forces, bases and support facilities there. Technologically, using them that way poses no problem: both cruise and Pershing missiles can be used over short as well as long distances. If the scenario is not a European battle but a direct confrontation between the major powers involving their intercontinental systems, then cruise and Pershing missiles in Western Europe would be available for use against the Soviet Union in accordance with US strategic warfare plans.

Following this line of reasoning, the *targeting policy* for the cruise and Pershing missiles will comprise alternative sets of targets, tailored to different war scenarios. In a European battle, the missiles are likely to be directed at Eastern Europe; in a strategic exchange, at the Soviet Union.[14] They would figure in US strategic planning as well as in NATO's nuclear warfare plans. A historical precedent for such a solution might be recalled: the Mace B 2 500-km range cruise missiles deployed in FR Germany in the period 1962–69 were reportedly targeted by SACEUR against East European countries and by SAC against the Soviet Union [30]. If there is one reload per SS-20 launcher, and one of the missiles is a single-RV version with intercontinental range, some targeting flexibility of a similar kind may exist also for the Soviet Union.

[14] Pershing II targeting lists may, for instance, include targets allotted to the Pershing IAs that are replaced.

By fixing alternative sets of targets, a targeting policy for the new missiles may therefore be agreed upon. But a mutually agreed *employment policy* (specifying under which circumstances and for what purposes the missiles should be used) is much less likely to see the light of day. That has always been provisional, procedural or unfinished business with NATO's Nuclear Planning Group (NPG).

While politically related to questions of alliance cohesion, the main US military rationale for the new missiles therefore appears to be strategic. The Pershing II will be one of the most capable counterforce weapons in the US arsenal should it ever be deployed in Europe: the RADAG system allows it to home on to virtually any kind of fixed target; no effective defence against it exists; and the flight time would be 12 minutes or less, depending on the distance to target. The flight time for Soviet LRTN missiles is correspondingly short. At worst, this may become another argument for adopting launch-on-warning strategies, which would increase the danger of nuclear war by accident.

The modernization decision and the countervailing strategy

The preparations for the 12 December decision and the elaboration of the countervailing strategy, laid out in Presidential Directive-59 (PD-59, of July 1980), took place in parallel. Little is known about the relationship between the two.

Generally, the countervailing strategy seems to have taken existing US strategy a few steps further, representing continuity rather than radical new departures. It requires that US forces not only maintain the capability for assured destruction of the Soviet Union, but also have "the capability for flexible, controlled retaliation against a full range of targets for any attack at any level" and, in so doing, confirms the changing direction of strategic doctrine that has evolved over a number of years [31].

PD-59 placed stronger emphasis on the capability to destroy military and political C^3I facilities, and raised a great demand for new weapons capable of knocking out hard targets. Cruise and Pershing missiles were technologically close at hand; they are both suitable for hard-target missions, and the Pershing II is ideal for use against time-urgent targets. Speeding up cruise and Pershing programmes was therefore a logical response to the new requirements as defined by the countervailing strategy. In this way, two endeavours pursued for their own reasons may, in effect, have become mutually reinforcing. This may partly explain why the Carter Administration—which initially stressed that the United States had more than a sufficient number of intercontinental systems to cope with the Soviet theatre threat—switched to a position of strong support for new deployments in Europe. The breakthrough for the introduction of the new missiles on European soil apparently came at the four-power summit meeting at Guadeloupe in January 1979, on President Carter's proposal.

In reviewing the capabilities of existing US intercontinental systems in relation to the requirements defined by the countervailing strategy, the Comptroller General concludes that only bomber-delivered weapons—from the FB-111A bomber—have the necessary combination of yield and accuracy to destroy efficiently super-hardened targets while limiting collateral damage. Of the near-term programmes, only ALCMs have the required accuracy/yield combination [31]. Although not reviewed in the report, it goes without saying that the GLCM is similarly suitable for such missions. However, for limited nuclear strikes, weapons which can destroy assigned targets with certainty while minimizing collateral targets are required: because it is vulnerable, the cruise missile is less well suited for this purpose. The Pershing, on the other hand, has superior characteristics for limited strikes, almost regardless of the hardening of the target, and is ideal against time-urgent targets.[15] The weapons planned for Europe therefore fit the requirements of PD-59 very well.

However, in relation to the total demand for new weapons raised by the countervailing strategy, the deployment decision of 12 December constitutes only a partial, small-scale response. While taking the new missiles into account, the strategic planners might, furthermore, not want to rely very much on them—for any single mission—since there are alternative, non-strategic uses for the same weapons as well.

VI. Divergent interests across the Atlantic: European concerns

In 1977 Chancellor Schmidt called attention to the implications for alliance strategy of the combination of parity in intercontinental systems and disparities in the European region [33]. After the agreement to modernize had been reached at Guadeloupe, the HLG suggested that somewhere between 200 and 600 missiles should be deployed: fewer than 200 would be of too little concern to the Soviet Union, while more than 600 was found too threatening (to the Soviet Union) by many participants in the Group. The 108/464 mix was finally picked by the United States, and formally adopted by the NPG meeting in November 1979 [34]. The

[15] "Potential Pershing II targets include: hardened and soft missile sites; airfields; naval bases; nuclear, biological and chemical storage sites; command and control centers; headquarters; rail yards; road networks/choke points; ammunition and petroleum storage facilities; troop concentrations and facilities; and dam/locks. Pershing II is particularly effective against hard point and underground targets . . ." [32].

West German government demanded that the final NATO decision be unanimous, and that at least one other continental European non-nuclear weapon country also accept the stationing of new missiles (the principle of non-singularity). Being particularly sensitive to political blame, it wanted the new missiles in FR Germany to be under complete US control. Italy, on the other hand, has shown interest in a two-key system similar to the one that existed for the Jupiter missiles 20 years ago [35].

The other track of NATO's 12 December decision—the invitation to negotiate, with the 572 missiles providing bargaining leverage—was promoted mainly by West European countries, particularly by some of the smaller ones. While important European government segments still associate themselves with the original rationale for modernization, the priorities differed on the two sides of the Atlantic—and increasingly so with the advent of the new US Administration. West European countries are fundamentally interested in coming to grips with the Soviet LRTNF threat, and public opinion against the deployment of cruise and Pershing missiles is very strong [22a]. The United States, on the other hand, seems much more bent on deploying the new missiles.

Once more, weapons that were in large measure justified as bargaining chips may therefore prove difficult to get rid of, although the original military-strategic rationale for deploying them has been increasingly questioned since the 12 December decision was taken. In some ways, the new missiles clearly make a difficult situation even worse for Western Europe; they burden the host countries with a number of high-priority nuclear weapon targets, and they will draw Western Europe into any strategic war between the two great powers. Even today, it is very likely that West European countries will be involved in such a war. With the new missiles on their soil, that likelihood approaches certainty. Equally important but often neglected in the public debate, the East European countries will also have to pay. In a European battle, they are likely to be the nuclear victims of the cruise and Pershing missiles, while the West Europeans face destruction from the SS-20s.

The modernization programme was an effort to give more credibility to the nuclear umbrella, or at least maintain ambiguity in relation to questions of nuclear strategy, glossing over diverging national interests across the Atlantic. To make the US commitment more visible and thereby convincing, deployment in a land-based mode was preferred. Far from reassuring the Europeans, however, that visibility made strong public opposition even more powerful than it would have been had the missiles been deployed at sea or in another, less transparent mode.

In addition, there is the fear that ever more effective war-fighting weapons may be introduced in a major-power competition which is not of European making, but which may make Europeans the main losers.

In the East, the Soviet SS-20 programme goes on: by mid-1981, the United States reported that sites had been prepared for an additional 65 launchers [36]. While the SS-20 currently has a CEP of 400 metres, it may approach 200 metres after perfection of the inertial guidance techniques. Should arms limitation efforts not succeed, a new, lighter, more mobile and terminally guided Soviet LRTN missile may, furthermore, appear towards the end of this decade. In the West, the range of the Pershing can be extended even farther, and it is claimed that the technological basis exists for installing a terminal guidance system which would make it effective against mobile targets as well—such as SS-20s on the move [37]. Greater numbers have also been considered.[16] However, for the time being, all proposals to exceed the quantitative and qualitative levels that were defined in 1979 are in abeyance: repeating them would only make it more difficult for the European governments to stand by the 12 December decision. As for cruise missiles, ALCM and SLCM programmes should, moreover, obviate whatever interest remains for more GLCMs.

For the first time in 20 years, there is today strong public opinion in Europe asking for the reduction or elimination of nuclear weapons from the continent. Discussions and public manifestations seem more intense and wide-ranging than they have since nuclear weapons came to Europe. The outcome of this debate—which has been called 'a battle for the soul of Europe'—may significantly affect the political orientation of West European countries and the shape of their defence. While the USA certainly believes that the Soviet Union poses the greatest threat to its security, many West Europeans see the US–Soviet conflict as the primary threat to them.

VII. Approaches to arms limitation

Initial positions

In April 1979 NATO established a Special Group (later the Special Consultative Group, SCG) to study the arms control implications of the emerging modernization decision. The Group took as its starting-point the need for NATO to deploy new systems—the work of the HLG being the basic frame of reference—and that arms control negotiations should be complementary to rather than a substitute for modernization. The Group also agreed that the negotiations should be conducted within

[16] In the beginning of 1979, a US Defense Nuclear Agency study came up with military requirements for 1 500 warheads. This was quickly discarded as politically unfeasible. During 1980, proposals for more than 572 missiles were aired once more [38].

the framework of SALT III, and that the USA should seek equality in ceilings and rights, even if the West did not intend to exercise such rights. To begin with, the principal negotiating objective would be to reduce the deployment of SS-20s, and to ensure the complete retirement of SS-4 and SS-5 missiles. The Western system negotiable at this stage would be the Pershing IIs and GLCMs. These land-based missile systems were to be subject to global limitations as well as regional subceilings.[17]

On 6 October 1979, the Soviet Union offered to negotiate on the condition that NATO would defer its ensuing decision to deploy new missiles. NATO rejected the offer, and the Soviet Union later declared that NATO's 12 December decision had destroyed any possibility for negotiations. That possibility received another setback when the Soviet Union intervened in Afghanistan. In response, President Carter asked the Senate to suspend consideration of the SALT II Treaty, bringing US–Soviet arms control negotiations to a full halt. The deadlock was broken by Chancellor Schmidt's visit to Moscow on 30 June–1 July 1980, when President Brezhnev declared that the Soviet Union was ready to negotiate even before the US ratification of SALT II, but that any resulting agreement could take effect only after US ratification. Furthermore, the missile question had to be discussed "simultaneously and in organic connection with the question of American forward-based nuclear means" [39].

In August 1980, Brezhnev sent a letter to President Carter and other Western leaders denouncing US reluctance to begin LRTNF negotiations. A month later, Secretary of State Muskie announced that the United States and the Soviet Union would open talks in Geneva. However, the first, preliminary round, which started on 16 October, quickly led nowhere, with sharp disagreement over which systems to include in the negotiations.

The United States presented the NATO position as agreed by the Special Group, emphasizing that the negotiations should be a step-by-step process, beginning with narrow and selective areas (i.e., land-based missiles) on the grounds that a comprehensive approach would raise a number of difficulties and complexities, minimizing the chance of progress. Therefore, the somewhat less urgent aircraft issues were to be considered at a later stage. The United States proposed that the counting unit should be warheads on launchers.

The Soviet Union held the view that if all LRTNFs are taken into account, a balance exists in Europe. A broad range of NATO nuclear capabilities were mentioned as suitable for use against targets on Soviet territory, and relevant to the overall balance. The Soviet negotiators

[17] Attempts to limit the Backfire bomber should be made in the context of SALT.

stressed the desirability of freezing the balance at existing levels, and proposed launchers as counting units.

In his speech to the 23rd Party Congress in February 1981, President Brezhnev proposed a moratorium on the deployment of LRTNFs. The moratorium would enter into force the moment negotiations began on the subject, and would operate until a permanent treaty was concluded. Both sides were expected to stop all preparations for the deployment of additional weapons. It was later indicated that the proposal did not require a halt in the production of missiles, since missile production could not be verified. Thus, US production of Pershing IIs and GLCMs might continue, while preparation of sites in Europe would presumably have to be stopped [40]. The proposal was rejected by the West as it would freeze a situation which was seen to be grossly unfavourable to NATO, and for fear that it would leave SS-20s east of the Urals untouched.

The policy of the new Reagan Administration towards the Soviet Union in general, and arms control in particular, suggested to many Europeans that the United States would not be serious about nuclear arms reductions. The Administration announced that it would undertake a comprehensive strategic review, and develop its arms control approach from there. Noting the general pre-conditions that were elaborated—negotiating leverage through arms programmes of unprecedented magnitude, linkage and new verification requirements—European worries persisted.

While both powers have stated that they will refrain from acting contrary to the provisions of SALT II until further notice, the United States made it clear that it would demand very substantial amendments. Provided that no major change occurs in the international environment, the US Administration has indicated that it will be prepared to resume the strategic arms reduction talks (START, an acronym introduced by the USA) in 1982. However, the future of these talks—of fundamental importance for LRTNF negotiations—seems highly uncertain.

To some extent, initial concerns over US intentions were ameliorated by the announcement at NATO's ministerial meeting on 4–5 May 1981 that the United States would "begin negotiations with the Soviet Union on TNF arms control within the SALT framework by the end of the year". Meeting at the United Nations on 24 September, Foreign Ministers Haig and Gromyko agreed to open negotiations in Geneva on 30 November 1981.

Overtures to Geneva

During November 1981, both the United States and the Soviet Union made far-reaching proposals for the reduction of nuclear arms which were very much addressed to public opinion in Europe. The anti-nuclear movement has become a major factor affecting the course of negotiations, so both sides evidently felt the need to please public opinion and to show that nuclear disarmament is a high-priority item on their foreign policy agenda.

In his speech at the National Press Club on 18 November 1981, President Reagan said that "the United States is prepared to cancel its deployment of Pershing II and ground launched cruise missiles if the Soviets will dismantle their SS-20, SS-4 and SS-5 missiles". This 'zero-option' proposal implied that *all* Soviet missiles of these types must be dismantled regardless of their location, including those deployed in Eastern Siberia.

In an address in Bonn on 24 November, President Brezhnev took the Soviet moratorium proposal a step further. Given agreement on a moratorium, the Soviet Union would not only be ready to halt the deployment of SS-20s, but would also be ready unilaterally to reduce the number of missiles in the European part of the USSR—"in other words engage in some anticipatory reductions moving to that lower level which could be agreed upon by the Soviet Union and the United States as a result of the talks". As part of an agreement, the Soviet Union would be prepared to make "reductions not of dozens, but of hundreds of individual weapons of this class". Brezhnev added that the Soviet Union was in favour of Europe finally becoming free of all nuclear weapons—of "all kinds of medium-range nuclear systems directed towards Europe . . . as well as of tactical weapons. That would be a real 'zero option' ".

Thus, under the influence of public opinion, both sides have adopted a declaratory policy which raises both the priority of arms negotiations and the ambition of achieving substantial reductions. Still, odds seem to be against a rapid turn of events: the general state of East–West relations and the domestic interests that influence negotiating positions are not conducive to radical departures.

Main issues

The SALT connection. The parties agree that the negotiations must be connected with a new round of negotiations on intercontinental strategic systems (often referred to as 'central systems').

The reduction of LRTNFs will certainly lose much of its significance if intercontinental systems are allowed to increase unchecked. Intercontinental weapons can be used over shorter distances as well, so if there is vastly more than enough for strategic deterrence, there is enough for regional assignments, too. SALT-accountable forces have been targeted on Europe in the past—some of them still are—and technically there is nothing to prevent this from happening on an even larger scale in the future. All the targets that can be struck by new Soviet and US theatre systems can be, or are, targeted by central systems as well.

Thus, while the renunciation of GLCMs would be a great relief for many Europeans, seen in the larger strategic context it would be of little significance if ALCMs and SLCMs are left unrestrained. In the short term, the wide dispersal of thousands of cruise missiles may appear very attractive to the United States. Any ship or submarine in the Atlantic, Pacific or Indian Ocean that may carry cruise missiles would be a potential threat to targets in the Soviet Union or to its allies. In the longer term, however, limitations on SLCMs are likely to be in the interest of the United States as well, since the Soviet Union may catch up and threaten US territory from the long Atlantic and Pacific shores. In this regard, the potential negative feed-back of an ambitious SLCM programme has been compared with the long-term consequences of the decision to MIRV intercontinental missiles. Much the same goes for ALCMs.

Another reason for linking LRTNF negotiations to SALT is that for obvious geographical reasons, global and regional parity between the two major powers cannot exist at the same time. LRTNFs in Western Europe or elsewhere on the Eurasian periphery can reach the Soviet Union, while the MIRVed SS-20s and other Soviet weapons in this category cannot reach the United States. The only regional level which is compatible with overall parity in strategic systems is defined by the figure zero. A zero solution making the deployment of cruise and Pershing missiles in Western Europe superfluous is therefore important not only for the security of European states, but also in the wider sense of facilitating US–Soviet strategic arms limitation efforts in the future.[18]

Experience suggests that negotiations which stop after agreement on some particular category of weapon has been reached in the long run prove futile, because the parties might begin to expand other forces not covered by the partial agreement. It is therefore important to see the LRTNF negotiations as the beginning of a long process which would soon overlap with strategic arms negotiations, and also lead on to systems of shorter range, that is, expand both up and down the ladder. Seven years of SALT negotiations have also shown that agreements must be more quickly negotiated than in the past.

[18] It should be noted that the strategic parity problem is particularly sensitive in relation to US LRTNFs, both because of the qualitative characteristics of the new missiles and because French and British missiles would not necessarily be fired in a direct exchange between the two major powers. The likelihood of staying outside a nuclear war between the two major powers may be very small, but in some scenarios, French and even British authorities may withhold the weapons in an effort to keep their countries out of the warfare.

Geographical domain

Press reports have indicated that the US Administration has dropped the idea of a European regional subceiling, focusing on global limitations instead. However, the only joint Western position that had been declared by the end of 1981 was that limitations should apply both world-wide and at the regional level.

The United States calls the Geneva talks "Intermediate Nuclear Force Negotiations", while the Soviet Union calls them "Talks on the Reduction of Nuclear Arms in Europe". The titles are indicative of a difference in geographical emphasis, the Soviet focus clearly being in Europe. The precise ramifications of this regional emphasis, for instance in relation to SS-20s deployed behind the Urals but within striking range of Europe, appear open to negotiation. Nor is it known whether, or under what conditions, the Soviet Union might eventually be willing to contemplate global ceilings.

In NATO deliberations, the inclusion of SS-20s on the Asian side of the Urals has been more of a European than a US demand. A case for taking even missiles deployed close to the Chinese border into account can be construed on the grounds that the SS-20 is mobile, and that with a single warhead targets in NATO Europe may be within reach. However, the missiles in the Far East are in all likelihood aimed at China, Japan and targets on other Asian territories, so to bring them into the European calculations is far-fetched.

While asking for the elimination of all Soviet LRTN missiles, President Reagan's zero proposal leaves the British, French and Chinese forces aside. This is rather extreme, because it seems to suggest that the Soviet Union has no regional security requirements in relation to the other three nuclear powers on the Eurasian continent.

The scope

While the initial US emphasis will be on the missiles covered by the zero proposal, the Reagan Administration reserves the right to seek limitations on Soviet SS-22 and SS-23 missiles to avoid circumvention of an LRTNF agreement. If deployed in sufficient numbers and moved forward on WTO territory, it is claimed that the SS-22 can cover about 85 per cent of the NATO targets assigned to SS-20s, and the SS-23 as many as 50 per cent [41]. Real negotiations on systems with a range below 1 000 km which, on the Western side, would have to comprise Pershing IA missiles may, however, be deferred to a later stage. Similarly, the West may still want to defer the aircraft issue until agreement has been reached on land-based missiles, although the position on aircraft appeared open to debate when the talks recessed on 17 December 1981.

The Soviet Union maintains that, overall, an approximate balance exists in the European theatre, with each side having an advantage in certain categories. It will almost certainly argue that any reduction should preserve the balance of forces that currently exists, and that the negotiating approach must therefore be a comprehensive one. Thus, by renunciation of all LRTNFs directed at Europe, the Soviet Union understands the renunciation of LRTN aircraft as well as British and French missile forces in this category. Included are US forward-based F-111s and F-4s, carrier-based A-6s and A-7s within striking range of Europe, and FB-111s based in the USA but intended for use in Europe.

The SS-20s deployed near the Urals are in a swing position between Europe and China, and some of them are likely to be targeted on the Middle East as well. However, they are all capable of hitting Western Europe, so by the criterion of striking range, they may all be included in the European calculations. Any partial inclusion of SS-20s in this area would, furthermore, seem arbitrary (and impossible to verify). On the Western side, some account has to be taken of British and French missiles in addition to the GLCMs and Pershings. The fact that France and the UK are not willing to take part in the negotiations is not decisive in this connection. Their forces can nevertheless be taken into consideration by allowing Soviet forces to vary correspondingly [5]. French and British missiles are, after all, directed at the Soviet Union, so for the negotiations to reflect military realities, some allowance has to be made for these forces even if they are not formally counted in the final balance.

SALT-accountable forces assigned to European missions—SS-11s, SS-19s, Yankee-class submarines which can use their missiles against Europe while in transit to and from stations near the east coast of the United States or from the Barents Sea, and the 400 Poseidon warheads assigned to SACEUR—need not enter the LRTNF calculations. Soviet SS-N-5 missiles on board Golf II-class submarines are not covered by SALT, and are treated by both sides as LRTNFs.

It is important that the terms of the negotiations be as simple as possible. With a degree of complexity similar to the Vienna M(B)FR negotiations (on mutual (balanced) force reductions), the negotiations are likely to be drawn out and inconclusive and, in effect, counter-productive. To begin with, a strong case can therefore be made for addressing the most urgent problems, namely the build-up of missiles, and leaving more complex issues such as LRTN aircraft aside. There is nothing inherently wrong with partial limitations, provided that they curb or reduce real threats and that they are not circumvented.

Also, for the Soviet Union, three factors would actually advise against the inclusion of bombers in the first phase of the negotiations. First, the complexity of the issue: the combat radius of aircraft depends on flight profile, speed, evasion of enemy air defences, payload, in-flight refuelling possibilities and availability of airfields, and these are seldom fixed quantities. Therefore, the inclusion of aircraft may well run counter to the strong Soviet interest in negotiating the non-deployment of cruise and Pershing missiles before the end of 1983. Second, the number of primary LRTN aircraft seems significantly higher on the Soviet side, both world-wide and for the European region. Third, improvements in Soviet air defence systems have reduced the penetrability of ageing Western air-craft, which for this and other reasons have become somewhat less of a threat over the past 15 years.

Generally, the Soviet Union nevertheless seems to prefer a comprehensive approach. This is a logical consequence of the view that an approximate balance currently exists in the European region. Equally important, a comprehensive approach makes sense because of perceived US efforts to gain some kind of military superiority. Superiority is not compatible with arms limitation agreements across the board, but only with partial agreements in areas not designated for achievement of upper edges. As that is the perceived context, the Soviet Union is likely to turn a sceptical eye on US proposals for narrow deals. Finally, the removal of US forward-based systems from Western Europe has been an important Soviet foreign policy objective for a long time, and still is.

However, this is not to say that the Soviet Union will necessarily insist on including LRTN aircraft in the first phase of the negotiations. A compromise might be struck between the quest for a comprehensive negotiation and for expeditious treatment of urgent missile issues, leading to a staged but integral process where the aircraft sector is brought in at a later phase.[19] For the United States, it is difficult to see how it could insist on broadening the scope of the negotiations to include SS-22s and SS-23s while continuing to deny the inclusion of forward-based aircraft.

Unit of account

While the Carter Administration had proposed warheads on launchers, the Reagan Administration proposed warheads on missiles as units of account, including reload missiles as well as those on launchers. The Soviet Union proposed launchers, in conformity with previous SALT practice.

Basically, there are three possibilities: to count launchers, missiles or warheads. There are arguments for all of them. Launchers are easiest to verify. Missiles make sense because there are four missiles per GLCM

[19] Soviet agreement to such an approach was indicated by Chancellor Schmidt in a speech before the *Bundestag* on 3 December 1981.

launcher, and reload and refire possibilities for ballistic missile launchers. However, missiles are hard to verify, and to date, none of the parties has shown much interest in making missiles the primary counting unit. In a way, warheads would be the best units of account because it is the warhead that kills, not the launcher or the missile. Warheads are much emphasized in the West because the SS-20 has been tested in a MIRV mode, while cruise and Pershing missiles carry single warheads only. However, if French and British forces are included, MIRVing of SLBMs may make warhead counting a dubious proposition for the West.

The US position therefore seems to be a maximalist stand premised on the exclusion of French and British forces. The verification of warheads on missiles is an extremely ambitious proposition, and the possibilities for verification are an unavoidable factor in the choice of counting unit. From that point of view, launchers would undoubtedly be the preferred alternative.

Numbers

President Reagan's 'zero option' was designated to be the US negotiating objective, and the US delegation reportedly had no fall-back position for the first round of the negotiations [41]. The even more encompassing zero option mentioned by the Soviet Union seemed to be a public relations counter to the Reagan proposal rather than a concrete negotiating objective: for instance, while British and French forces may be taken into account, their size cannot be determined by the two major powers, even less negotiated away. The main Soviet objective is undoubtedly to avoid the deployment of new US missiles in Western Europe.

In addition to its non-consideration of Soviet regional security requirements, the seriousness of Reagan's proposal can also be questioned on the ground that it is inconsistent with the original NATO rationale for the modernization decision. That decision was allegedly based on the judgement that new systems were required to sustain the doctrine of flexible response and enhance the credibility of the US nuclear umbrella over Western Europe. NATO had to modernize primarily because of its own force requirements, while the logic of the zero option suggests that the need to modernize would disappear with the SS-20s. The contradiction is obvious, but the problem is mainly for West European governments to sort out: following from the doctrine assessment made in the previous section, no technological fix can re-establish the coupling to US intercontinental forces that once existed, and the United States pursues the modernization programme for other reasons.

The prominence which the Reagan proposal gives to the figure zero may be taken as a concession to public opinion in Europe, because it holds out the possibility of avoiding the deployment of GLCMs and Pershings. Negotiating positions which are unrealistic in relation to this demand are likely to generate more public opposition than support. On the other hand, it can also be seen as a political tactic for making the Soviet Union responsible for new deployments: unless the Soviet Union does away with its SS-20s, NATO has to move ahead.

The threat assessment and functional requirements studies initiated by the Reagan Administration through the HLG can be interpreted in the latter direction. The threat assessment study emphasized both the speed and scope of Soviet TNF modernization, and the functional equipment study reconfirmed the need for both GLCMs and Pershings. For West European governments, 572 remains a definite high end of the modernization effort. The United States may, however, use the size of the Soviet TNF programme to support the view that 572 is at the low end of NATO requirements. This can, in turn, influence the number of NATO LRTNFs that it is willing to negotiate and that European governments would, in the end, go along with [24].

The Soviet offer to engage in some "anticipatory (missile) reductions" was a move in the right direction. International negotiations would have a better chance of success if accompanied by unilateral steps down instead of up. Apart from presupposing agreement on a moratorium, however, the offer must be seen against the background of existing Soviet predominance in missile systems, and the growing obsolescence of remaining SS-4s and SS-5s. When these missiles were deployed 20 years ago, they compensated for a clearly inferior position in intercontinental systems: this particular justification for Soviet LRTNFs no longer exists. The peoples of Europe are therefore entitled to expect substantial Soviet reductions in return for a cancellation of the Western modernization plan.

NATO's 12 December decision asked the United States to seek *de jure* equality in rights and ceilings. The Special Group had recommended it, but on the understanding that the West did not have to exercise such rights; the rationale for the modernization plan as developed by the HLG did not presuppose equality. However, the politics, psychology and experience of arms control strongly indicate that once this principle is established and a certain level agreed upon, no party will be satisfied with staying far below that limit.

Verification

The US Administration does not consider national technical means of verification to be adequate for an LRTNF agreement. More information has to be elicited from the Soviet Union: a radical improvement in Soviet willingness to provide data for the verification of future agreements is required. The Soviet Union emphasizes that national means have proved satisfactory in the past, that the effectiveness of these means is continually being improved, and that they should therefore have priority also in the future. However, also in the Soviet view, other forms of verification can be developed "if mutual trust is achieved" [42].

The introduction of cruise missiles makes it very hard to verify nuclear force deployments. There are different opinions on the verifiability of GLCMs, but less so regarding SLCMs: should the current plans for the wide dispersal of SLCMs be implemented, effective verification may well become impossible. Proliferating cruise missiles in an international atmosphere of deep distrust therefore raises unprecedented demands for ingenuity in the field of verification techniques. Substantial limitations on this technology, on SLCMs in particular, may therefore be of fundamental significance for the future of arms control.

Finally, it should be noted that zero-level agreements, prohibiting certain categories of weapons altogether, are by far the easiest to verify. The parties would then be expected to close down the factories; training for the weapons would not be justified; and there would be no weapon flight-tests. From a verification point of view, the difference between zero and one is salient, and that may go for cruise missiles more than for any other weapon.

In the autumn of 1981, the Soviet Union for the first time published figures and other information on its LRTNFs. This is a most welcome development. However, traditional Soviet secretiveness will not be abandoned overnight, and the United States seems to press for verification procedures that the Soviet Union is unlikely to accept.

Linkage

The United States takes the general view that there can be no arms control agreement without linkage. According to Secretary Haig, "we have learned that Soviet–American agreements, even in strategic arms control, will not survive Soviet threats to the overall military balance or Soviet encroachment upon our strategic interests in critical regions of the world. Linkage is not a theory: it is a fact of life that we overlook at our peril" [43]. Thus, Soviet concessions in places like Kampuchea and Afghanistan have been mentioned as pre-conditions for agreements on arms limitation. The West European allies are less prone to pursue linkage politics, especially in relation to the Geneva negotiations, where domestic stakes in a successful outcome run so high.

Linkage tends to enhance the prevailing trends in international affairs: in the first half of the 1970s, linkage politics was a deliberate strategy for the promotion of East–West co-operation and detente, whereas in recent years it has made tense US–Soviet relations even more intractable. Today, it turns arms limitation into a reward for good behaviour in other fields: if you step down here and here, we offer you mutual arms limitation in return. This is certainly hard to accept for the adversary, and hard to justify to security-minded constituencies in Europe. In reality, arms limitation and disarmament are in the common interest, both of the great powers and of the other nations of the world. In US politics, however, linkage is deeply rooted and difficult to get round.

An alternative approach: unilateral, reciprocal action

Over the past five years, the military preparedness to produce and deploy new LRTNFs has not been matched by political readiness to seek arms limitations. To a large extent, this period also has a history of unfortunate sequences and lost opportunities.

The negotiations that began on 30 November 1981 deserve support as long as they have a fair chance of succeeding. However, more than two years have passed since the 'dual track' decision was made, and the general prospects for nuclear arms limitations are bleak.

In Geneva, the parties do not agree on what to call the negotiations, what to negotiate, or what to count. They differ on geographical coverage, and there is a public relations battle over figures and how to define the military balance. Verification has once again become a controversial issue, and cruise missile technology poses very difficult verification problems.

If the negotiations become deadlocked, readiness to pursue an alternative course of unilateral, reciprocal action therefore seems important. Reciprocity can be achieved through tacit understanding, meaning East–West consultations to co-ordinate the moves undertaken by each side. That could make unilateral action more acceptable at home, and therefore easier to decide and implement.

For this approach to be pursued, the USSR should take the lead together with some key West European countries. It might be recalled that 10 years ago, *Ostpolitik* was largely pursued by FR Germany and France, with a number of more or less sleeping partners elsewhere in Europe. This time, the elaboration of a tacit understanding for reciprocal, unilateral action also depends on the right initiatives by a proper combination of countries. Again, FR Germany is a country of critical importance. The Soviet–West German agreement to consult regularly about nuclear weapons in Europe during the course of the Geneva talks is an interesting and potentially significant development, also from the point of view of readiness for unilateral, reciprocal action.

Today, European governments are urged to find a way out of the dilemmas posed by theatre nuclear weapons. Public opinion—stronger than at any time since World War II—demands a radical departure.

It would be most unfortunate if the governments of European countries were to become preoccupied with resisting and opposing these movements. Instead, they ought to seize the opportunity to reassess where we stand, approach the fundamental dilemmas of European security with the necessary vigour to improve our predicament, and give constructive direction to public activity in this field. Until recently, European leaders did not have the necessary public support to take such action, even if the desire was there. They were forced to live with the flaws and dilemmas. Today, the situation is different.

References

1. US Air Force, SAC, HQ, Office of the Historian, *Development of Strategic Air Command 1946–1976* (USAF, Washington, D.C., 1976), p. 5.
2. Haglund, C., *Tillbakablick på Utvecklingen av Ballistiska Robotar i USA samt Systembeskrivningar*, FOA 2 Rapport A 2552-D9, F5 (National Defence Research Institute, Stockholm, Sweden, 1972).
3. SIPRI, *Tactical Nuclear Weapons: European Perspectives* (Taylor & Francis, London, 1978), p. 125.
4. Armacost, M. H., *The Politics of Weapons Innovation: The Thor–Jupiter Controversy* (Columbia University Press, New York and London, 1969).
5. Garthoff, R. L., 'Brezhnev's opening: the TNF tangle', *Foreign Policy*, No. 41, Winter 1980–81.
6. Richardson, D., 'World missile directory', *Flight International*, Vol. 119, No. 3760, 30 May 1981.
7. Berry, F. C., Jr., 'Pershing II—first step in NATO theater nuclear force modernization?', *International Defense Review*, Vol. 12, No. 8, 1979, pp. 1303–308.
8. Richardson, D., 'Pershing II—NATO's smart ballistic missile', *Flight International*, Vol. 120, No. 3770, 8 August 1981, pp. 431–34.
9. 'Pershing II, cruise missile production is hurried up', *Washington Post*, 29 November 1981.
10. US Congress, Senate, Committee on Armed Services, *Department of Defense Authorization for Appropriation for Fiscal Year 1981, Hearings* (US Government Printing Office, Washington, D.C., 1980), Part 2, p. 520.
11. US Comptroller General, *Most Critical Testing Still Lies Ahead for Missiles in Theater Nuclear Modernization* (MASAD-81-15), (General Accounting Office, Washington, D.C., March 1981).
12. Robinson, C. A., Jr., 'Decisions reached on nuclear weapons', *Aviation Week & Space Technology*, Vol. 115, No. 15, 21 October 1981, pp. 18–23.
13. 'Second-source contract expected on Tomahawk', *Aviation Week & Space Technology*, Vol. 115, No. 9, pp. 61, 63.
14. Carnesale, A., 'Reviving the ABM debate', *Arms Control Today*, Vol. 11, No. 4, April 1981, p. 2.
15. Dale, R., 'Five ways to ensure the survival of the U.S.', *Financial Times*, 3 October 1981.
16. *Air & Cosmos*, Vol. 19, No. 885, 12 December 1981, p. 2.
17. 'USAF updates nuclear capable F-111s', *Aviation Week & Space Technology*, Vol. 113, No. 1, 7 July 1980, pp. 48–49.

18. SIPRI, *World Armaments and Disarmament, SIPRI Yearbook 1979* (Taylor & Francis, London, 1979), pp. 227–28.
19. 'Fighters—improving the breed', *Flight International*, Vol. 120, No. 3783, 7 November 1981, pp. 1369–75.
20. 'Safety features of cruise missiles', *The Times*, 15 May 1980.
21. Robinson, C. A., Jr., 'USAF pushes production performance', *Aviation Week & Space Technology*, Vol. 114, No. 11, 16 March 1981, p. 50.
22. Sloan, S. R. and Goldman, S. D., 'NATO theatre nuclear forces, modernization and arms control', *Issue Brief* (IB81128), Congressional Research Service, 8 April 1981.
    (a) —, p. 6.
23. Corterier, P., 'Modernization of theatre nuclear forces and arms control', *NATO Review*, No. 4, 1981.
24. Lunn, S., 'LRTNF negotiations in Geneva: problems and prospects', background paper prepared for the Fifth Pugwash Workshop on Nuclear Forces in Europe, Geneva, 11–13 December 1981.
25. Senghaas, D., 'Questioning some premises of the current security debate in Europe', *Bulletin of Peace Proposals*, Vol. 12, No. 4, 1981.
26. Barnet, R., 'Why on earth would the Soviets invade Europe?', *Washington Post*, 22 November 1981.
27. Pincus, W., 'NATOs nuclear strategy devised in political vacuum', *Washington Post*, 15 November 1981.
28. SIPRI, *World Armaments and Disarmament, SIPRI Yearbook 1975* (Almqvist & Wiksell, Stockholm, 1975), pp. 41–46.
29. Opening address at the conference on 'NATO, The Next Thirty Years', Brussels, 1–3 September 1979.
30. Nerlich, U., 'Theatre nuclear forces in Europe: is NATO running out of options?', *Washington Quarterly*, Vol. 3, No. 1, Winter 1980.
31. US Comptroller General, *Countervailing Strategy Demands Revision of Strategic Force Acquisition Plans* (MASAD-81-35), (General Accounting Office, Washington, D.C., August 1981).
32. US Congress, House of Representatives, Subcommittee on the Department of Defense, *Department of Defense Appropriations for 1980, Hearings* (US Government Printing Office, Washington, D.C., 1979), Part 2, p. 863.
33. Schmidt, H., 'The 1977 Alastair Buchan Memorial Lecture', *Survival*, Vol. 20, No. 1, January–February 1978, pp. 2–10.
34. Pincus, W., 'Arms decision stirred storm around NATO', *Washington Post*, 18 November 1981.
35. Pincus, W., 'Allies split on touchy problem of who controls nuclear launches', *Washington Post*, 16 November 1981.
36. US Department of Defense, *Soviet Military Power* (US Government Printing Office, Washington, D.C., 1981), p. 27.
37. Moore, R. A., 'Theatre nuclear forces, thinking the unthinkable', *International Defense Review*, Vol. 14, No. 4, 1981.
38. Institute of American Relations, *Independence Through Military Strength* (IAR, Washington, D.C., 1980), p. 23.
39. Resolution adopted by the Politbureau of the Central Committee of the Soviet Communist Party, 4 July 1981, *Keesing's Contemporary Archives*, 19 September 1981, p. 30471.
40. 'Soviet expert on US presses arms talk call', *Baltimore Sun*, 17 March 1981.

41. Testimony of Richard Perle before the Senate Armed Services Committee on 1 December 1981, quoted in *Defense Daily*, 2 December 1981.
42. Brezhnev, L., in *Pravda* (and *Der Spiegel*), 3 November 1981.
43. Haig, A., Speech before the American Bar Association, New Orleans, in *Defense Daily*, 13 August 1981.

9. The CSCE and a European disarmament conference

The 35-nation Conference on Security and Co-operation in Europe (CSCE) met in Madrid on 11 November 1980 to review the implementation of the Helsinki Final Act. The negotiations in this forum have been devoted *inter alia* to the discussion of proposals for a European Disarmament Conference (EDC).[1] The Madrid meeting continued throughout 1981 and reconvened on 9 February 1982, after the New Year recess. By the end of 1981 the participants had still not agreed on a concluding document, which would establish a European disarmament conference.

I. Background

The idea of a European disarmament conference was much discussed internationally in the late 1970s, mainly because the failure of global disarmament efforts made regional initiatives more urgent. The talks held in Vienna between NATO and the Warsaw Treaty Organization (WTO) on troop reductions in Central Europe had been pursued since 1973, also without result. The European states which were not present at these talks—that is, France and the neutral and non-aligned European nations—emphasized the need for an all-European disarmament forum. There was also some support within NATO and the WTO for convening such a conference. The arms race in Europe was intensifying, with the prospect of further substantial additions to the stocks of conventional and nuclear weapons on both sides: something should be done to try to stop this development. The second follow-up meeting of the CSCE, which was to open in Madrid on 11 November 1980, was considered an appropriate forum for more detailed debate and for a decision on a European disarmament conference. But, since the signing of the Helsinki Final Act on 1 August 1975, the movement towards detente in Europe had been, if anything, reversed. The first follow-up meeting, held in Belgrade in 1977/78, had ended in failure.

Between Belgrade and Madrid such events as the continuing deployment of SS-20s, the decision on the future deployment of Pershing II and ground-launched cruise missiles (GLCMs) in Europe as well as the Soviet intervention in Afghanistan had further increased the tension between the

[1] See SIPRI's account of these discussions up to March 1981 in the *SIPRI Yearbook 1981*, chapter 17.

two major military blocs. Therefore, even before the meeting started, there was not much optimism about the outcome.

The first stage of the Madrid meeting coincided with the inauguration of the Reagan Administration and with increased tension in and around Poland. The harder line which the USA was taking towards the USSR, and the lengthy uncertainty surrounding the new US disarmament policy, clearly affected the tenor of the negotiations in the early months. However, the longer the meeting was permitted to drag on, the more difficult it became to bring it to an end, since no tangible positive result was in sight. None of the participating states would take the blame or responsibility for bringing about a final break-down. Consequently, the Madrid meeting, originally intended to end in the beginning of March 1981, was still in session in February 1982 (as this chapter goes to press). The faint hope seems to be that something might be achieved outside the meeting, for instance at the Geneva talks on theatre nuclear forces, which might give an impetus to the Madrid discussions and bring about some positive results.

The question of the early convening of an EDC has become the major issue at the meeting. Five delegations (from France, Poland, Romania, Sweden and Yugoslavia) have tabled proposals towards this end, with the French proposal in the main reflecting the Western position, and the Polish the Eastern.

II. The EDC proposals

Although all proposals seemed to aim at the same goal, the approaches were widely different, particularly those of the Polish and French proposals; these represented the main opposing positions and provided the frame-work for most of the ensuing debate. There were also proposals of some importance from eight neutral and non-aligned delegations for an enlargement of the confidence-building measures (CBMs) of the CSCE Final Act (see the *SIPRI Yearbook 1981*).

The Polish scheme suggested a step-by-step development from the same, existing CBMs—that is, voluntary-type—towards more complex and far-reaching measures of restraint and reduction of forces and armaments in Europe. It also assumed that any proposal within the scope of the conference and submitted by a participating state would be examined.

Western delegations rejected this all-embracing, unconditional approach, claiming that it would simply create a new arena for propagandist oratory and political declarations of intent rather than for serious negotiations. They also refused to build on the present CSCE CBMs, on the grounds that they had proved to be militarily insignificant and, in addition, had been unsatisfactorily implemented, to a great extent owing to shortcomings in the Final Act. The Western delegations were consequently also negative towards the proposals for second-generation CBMs which were submitted by the neutral and non-aligned delegations.

The neutral and non-aligned paper suggested *inter alia* that participating states should notify their major military manoeuvres and movements of troops exceeding a total of 18 000 men. One of the existing CBMs applies to manoeuvres of more than 25 000 troops. The prior notification of troop movements, as well as manoeuvres, and a lowering of the threshold to 18 000 would mean a considerable improvement.

Another new CBM suggested was to notify naval exercises in European waters involving major amphibious forces of more than 5 000 troops or 10 major amphibious vessels. Amphibious forces are typically designed for offensive purposes and surprise operations and their military significance is already recognized in the Helsinki Final Act. Prior notification of such exercises would have a considerable confidence-building effect. The term "European waters" was defined in the proposal as "the inner seas of Europe, i.e., the Baltic, the North Sea and the Black Sea, the Mediterranean and the ocean areas adjacent to the territorial waters of the European participating states".

Further, it was proposed that the potential confidence-building effect of increased openness in military matters should be recognized—particularly with regard to military expenditure.

The arguments for these improved and enlarged second-generation CBMs, on the lines of the voluntary CBMs in the Helsinki Final Act, were as follows: they would give new life to the original modest set of CBMs, which had been in force for seven years; and they could pave the way for more important decisions at a later stage. The more ambitious Western proposals for another, new type of CBM might well take years of negotiation. There was, after all, no strong pressure of public opinion in their favour—most people knew nothing at all either about existing CBMs or about any new proposals. The political atmosphere was much less favourable than it had been at the time of the Helsinki conference, when a much more modest set of proposals took some years to negotiate.

Because of the Western delegations' opposition, these important proposals were never discussed at the meeting and finally disappeared entirely from the draft concluding document which the same eight delegations tabled before the December 1981 recess.

The French proposal, like the Polish, suggested a step-by-step approach but emphasized the first stage: the adoption of a coherent system of new, not 'second generation', CBMs. Such a system was described by the delegation from the United Kingdom as "an arms control regime of openness" where regular information would be exchanged on all major military formations in Europe, from divisional level upwards; on the nature, designation and location of garrisons; and also on military movements, whether for exercises or for other reasons.

The French proposal tried to lay the ideological basis for such an ambitious project by demanding that four criteria should be agreed upon at the Madrid meeting before an EDC was convened: the new CBMs should be significant in military terms; they should be binding, not voluntary; there should be appropriate verification; and they should be applicable throughout Europe, from the Atlantic to the Urals. Depending on the results achieved at an EDC, a later CSCE follow-up meeting would examine how to continue "towards security and disarmament". Since, at the time of submitting the proposal, "disarmament" was almost a taboo word for some Western delegations (more armaments for catching up with the Soviet Union were considered the key towards redressing the European balance), there was no elaboration of the second, disarmament phase in the French proposal.

The first three criteria—military significance, binding obligations and appropriate verification—are likely to be accepted by all participating states. (What is to be understood by "appropriate" or "adequate" verification will, no doubt, remain a matter for lengthy discussion, as it has for years in other arms control contexts.) The area of application, however, has turned out to be the major controversial issue. The reason for this can be traced to the CSCE Final Act provisions for the prior notification of major military manoeuvres. There it was agreed that, whereas all other European states would notify such manoeuvres within their whole territory, for the Soviet Union (and Turkey) prior notification need be given only for manoeuvres which take place in an area within 250 kilometres of the frontiers which face other European states. This meant that about 80 per cent of the European part of the USSR was not included in the application of this measure, which had been designed as a modest first step to help prevent surprise attacks from areas near the borders of neighbouring states.

This exception from the CSCE "whole of Europe" concept had been a negotiation success for the Soviet Union, mainly because some Western states were then not particularly interested in the question. It was accepted at the time as striking a kind of geographical and strategic balance, since US and Canadian territories are not included in the area of application. But when far more important CBMs were being considered for the EDC agenda and when disarmament measures in Europe might later appear on that agenda, the area problem became much more important. The catchy French phrase "from the Atlantic to the Urals", originally coined by de Gaulle, was rejected by the Eastern states, which claimed that the Final Act area provisions had been accepted as a principle and should be valid also for other CBMs.

The long-drawn-out debate on the area of application, which is still the main stumbling-block preventing a decision about an EDC, has been going on now for over a year. In February 1981, at the Soviet Communist Party Congress, Leonid Brezhnev stated that the Soviet Union would be willing to apply CBMs to the entire European part of the USSR, provided that the Western states also extended the confidence zone accordingly.

The problem was then how to compensate for this Soviet concession "accordingly". A geographic approach would be to draw border lines or establish zones in the oceans and waters surrounding Europe, where military activities would be notified mainly along the same principles as in Europe itself. Another, functional, approach would be to select certain military activities outside the European territory, but connected with activities in Europe, for the application of any measures adopted. A third possibility—perhaps the most feasible—would be to combine geographical and functional elements.

In a neutral and non-aligned paper presented on 31 March 1981, CBMs—which had then been renamed CSBMs (confidence- and security-building measures)—were suggested to cover "the whole of Europe with the adjoining sea area and air space". This was, for different reasons, not agreeable to either the Eastern or the Western side.

In July, before the summer recess, the Western states were, however, reportedly willing to agree that the measures would be applicable to the whole continent of Europe, and also to the activities of forces operating in the adjoining sea area and airspace, insofar as these activities were an integral part of notifiable activities on the continent. This was, however, not accepted by the Eastern states.

Finally, just before the December 1981 recess, a compromise was suggested by the neutral and non-aligned states in a draft final document in which the Western text from July was supplemented with the idea that the necessary specifications of the area to be covered would be made in the negotiations on the confidence- and security-building measures at the disarmament conference itself, in the hope that by then there would be a better international climate.

The developments in Poland since December 1981, and the consequent US sanctions against Poland and the USSR, will not have made the problems in Madrid, or at any other comparable meeting, easier. A possible break-down of the CSCE would, however, be a very severe blow to the promotion of European security and co-operation. The unilateral and multilateral adoption and careful implementation of significant confidence- and security-building measures would provide convincing evidence that the intentions of the major powers were genuinely non-offensive and peaceful, which is, after all, what both sides repeatedly claim. Building confidence also requires that states abstain at the negotiating table from magnifying trivial matters into major national security concerns. Europe at present seems to be moving towards an intensified military confrontation. A European disarmament conference is badly needed, as a first step toward checking this process, and no effort should be spared in the attempts to bring such a conference about.

10. Nordic initiatives for a nuclear weapon-free zone in Europe

Square-bracketed numbers, thus [1], refer to the list of references on page 208.

I. Introduction

In 1961 Swedish Foreign Minister Undén suggested the creation of a 'club' of states obligated not to acquire nuclear weapons and not to accept deployment of nuclear weapons on their territories. In 1963 President Kekkonen of Finland adapted and confined Undén's idea to the Nordic region, proposing a Nordic nuclear weapon-free zone (NWFZ).[1]

The overriding concern behind the Kekkonen proposal was to keep the Nordic countries out of "the realm of speculation brought about by the development of nuclear strategy", and to maintain a state of low tension in the area. That same concern prompted a revised version of the proposal in 1978 [1], and has been an important impetus for the recent surge of interest in the zone issue in all the Nordic countries, precipitated by a programme declaration of the governing Labour Party in Norway.[2]

In the following sections, the main issues and problems connected with the creation of a Nordic NWFZ are discussed under 10 subject-headings.

II. Objectives

The overall objective of the Nordic NWFZ proposals is to strengthen the security of the countries in the region, and to stabilize relations between the big powers in this strategically important area.

The constellation of ground forces in northern Europe has remained stable for a number of years. Both Eastern and Western countries have shown restraint. However, military capabilities at sea and in the air are rapidly increasing in the region, threatening the security interests of all parties—Eastern, Western and neutral.

[1] In a letter to the Prime Minister of Norway of 8 January 1958, Soviet Prime Minister Bulganin mentioned the possibility of making northern Europe a zone free of nuclear weapons. In 1959, Prime Minister Khrushchev proposed a NWFZ in the Baltic area. The Polish disengagement proposals concerning Central Europe were more important for later Nordic initiatives; the Rapacki plan of 1957 was the first fully elaborated NWFZ proposal to be presented to the United Nations.

[2] The platform adopted by the Party Convention on 2–5 April 1981 reads: "The Labour Party will work for a nuclear weapon-free zone in the Nordic area as an element in the work to reduce nuclear weaponry in a larger European context".

More than two-thirds of the Soviet naval construction and repair facilities are located in the Baltic Sea, and the traffic through the Danish straits is therefore rather heavy. The Soviet Northern Fleet, home based on the Kola Peninsula, sustains the Soviet global military posture, and is an important source of reinforcement for conflict areas and battlefields in the Third World.

About 70 per cent of all the Soviet ballistic missile submarines (SSBNs) are in the Northern Fleet. Accordingly, the Norwegian and Barents Seas are a high-priority arena for US and British ASW (anti-submarine warfare) activities. Conventional and nuclear land-attack cruise missiles are planned for deployment on US attack submarines by 1982 and 1984, respectively, northern European waters being one of the likely deployment areas. The United States also plans to upgrade the presence of carrier groups in the North Atlantic and the Norwegian Sea during the 1980s, as one element of a comprehensive forward strategy to be enacted throughout the decade [2, 3].

Northern Europe is therefore an increasingly important *arena* of international rivalry, although it is not itself a *source* of major power conflict.

Against this background, a NWFZ may be an instrument by which the Nordic countries can exert some moderating influence on the military activities in their immediate surroundings. Any NWFZ in Europe would have the character of a buffer zone, and the elimination of nuclear weapons deployed in the vicinity of the Nordic countries and suitable for use against them would have to be part of the Nordic zone arrangement. However, for such deployment limitations to be realized, the major powers would have to see some common interest in avoiding tension in the area; they might then be interested in some zone design which serves that purpose.

III. Characteristics

There are three main characteristics of a NWFZ: non-possession, non-deployment and non-use of nuclear weapons. The non-possession requirement is already met by all the Nordic countries: they were among the first to ratify the Non-Proliferation Treaty (NPT). The non-deployment obligation, however, presents several difficulties for the two Nordic NATO members [4].

Norway and Denmark do not allow the deployment of nuclear weapons on their territories in time of peace. This is a unilateral measure of restraint; therefore, they are free to change policy at will, and options for the use of nuclear weapons on or from Danish and Norwegian territory have existed for years. Unlike the NPT commitment, this is a policy that can be changed

Figure 10.1. The northern European region

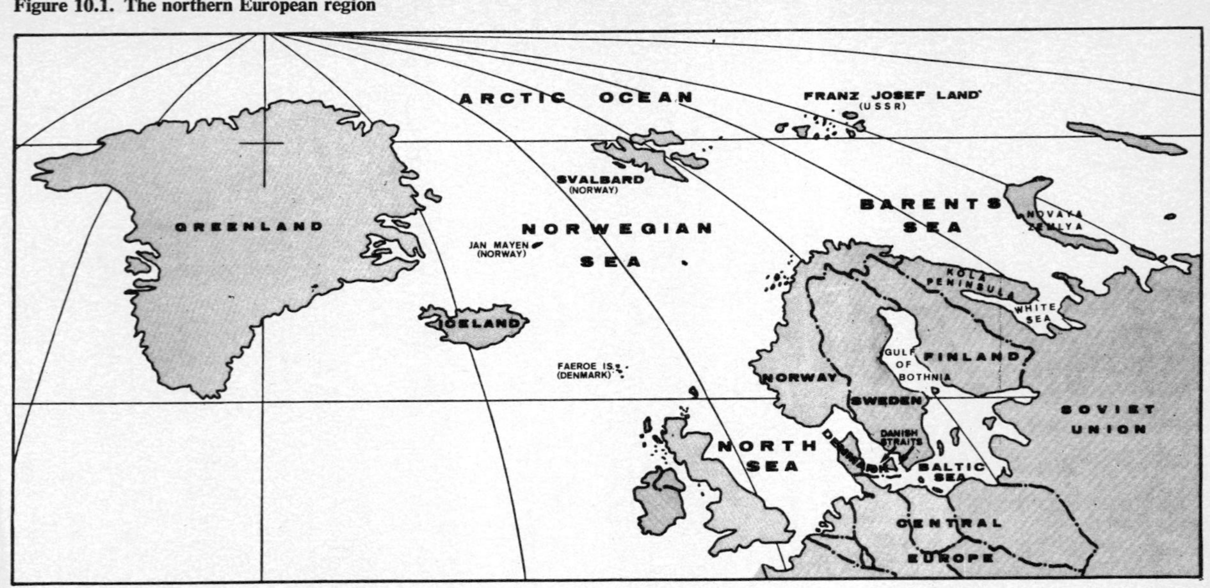

overnight. However, the broad consensus that has been formed around the non-deployment stand makes it hard for any government to back out of it under normal international conditions. Only a crisis could induce Norway and Denmark to ask for the transfer of nuclear weapons to their territories. Since the policy was instituted more than 20 years ago, technological developments have, moreover, rendered the exercise of the nuclear weapon option in time of crisis less important.

However, participation in a NWFZ would require an unqualified position against the deployment of nuclear weapons, applying in times of both war and peace, and embodied in an international legal instrument. While the policy of non-deployment in peace-time has never been challenged by other NATO members, non-deployment in wartime would impose a more substantial restraint on NATO nuclear planning for northern Europe. In important respects, Norway and Denmark would be decoupled from NATO's nuclear strategy, and their participation in NATO's military organization might have to be reconsidered also in other respects.

In the Final Document of the first UN Special Session devoted to Disarmament, held in 1978, the nuclear weapon states are called upon to respect the status of zones freely arrived at and to refrain from the use or threat of use of nuclear weapons against the states of the zone (so-called negative security assurances). In relation to the NWFZ established for Latin America by the 1967 Treaty of Tlatelolco, all nuclear weapon states have undertaken such obligations by ratification of an Additional Protocol to the Treaty, although with some important reservations [5]. A Nordic initiative for a NWFZ in Europe might follow that precedent, asking for negative assurances to be given in the same manner.

IV. The meaning of 'nuclear weapon-free'

'Nuclear weapons' usually means 'bombs and warheads'. By the established definition of a nuclear weapon-free zone, the prohibition applies to nuclear explosives only. It is in return for this prohibition that the nuclear weapon states are supposed to extend non-use assurances.

However, there are arguments for broadening the scope of a zone arrangement so as to prohibit other components of nuclear weapon systems as well.

It may seem artificial to single out bombs and warheads for exclusive attention and prohibition; rather, it could be argued that states from whose territories nuclear attack can be launched—because they have allowed nuclear explosive devices to be stationed on their soil or because they have permitted other vital components of nuclear weapons to be established on their territory—can only aspire to an assurance that they will not be subject to a *first* nuclear strike. Hosting important elements of nuclear weapon systems, they cannot be immune to a response in kind if a nuclear attack can be sustained from territory under their jurisdiction. After all, explosives are only one of the many necessary components of a nuclear weapon system. These components may figure on the nuclear targeting lists of adversary powers even if the territory on which they are located is declared nuclear weapon-free in the traditional sense.

If the NWFZ concept is limited to bombs and warheads, then there may be installations within the zone which could be used by another power in a nuclear attack, and which may figure on nuclear targeting lists, non-use guarantees notwithstanding. While there is no way of knowing that this is the case, military logic might indicate that it is, thereby detracting from the credibility of the guarantees. The established zone concept is clearly inadequate in this regard.

There are several examples of such installations presently in the Nordic area: navigation aids for submarines, communications-interception and direction-finding stations that can be used for target acquisition, and sonar arrays. The latter can be used by US Orion and British Nimrod aircraft carrying nuclear depth charges, and by attack submarines. However, these installations are all multi-purpose, and their actual significance for nuclear warfare can be disputed. It is often hard to determine whether a facility is an important part of a nuclear weapon system: this is a difficult task at any point in time, and the pace of military technological development makes it even more difficult to establish criteria for what is significant and what is not.

A zone arrangement must be perfectly clear as to rights and obligations: lack of clarity may lead to misunderstandings and suspicion, and guarantor states can make use of ambiguous provisions to exert pressure on member states. Clarity would appear to be an overriding concern. However, it is difficult to find an extended definition of denuclearized status which discriminates as clearly between things permitted and things prohibited as the distinction between presence and non-presence of explosives. This difficulty therefore amounts to a strong argument for sticking to the established meaning of 'nuclear weapon-free'. Should a country like Norway ever want to go beyond this and eliminate US or NATO-related facilities which may become nuclear targets in war, it could raise this question with other NATO members on a bilateral or alliance basis. In the NWFZ context, it would be another complication and, possibly, a major obstacle.

A zone arrangement implies, however, that all plans for the transfer of nuclear weapons to members of the zone must be scrapped. For instance, collocated operating bases (COBs) might be affected. The need for allied air support, essential for the defences of Norway and Denmark, must be made compatible with a credible non-nuclear status. This might be achieved either by changes in current agreements and practice, or by extended national verification rights, or by elements of both. At present, there are two Danish airfields in the COB programme and eight in Norway, in the total of some 70 for NATO Europe.

NATO members joining the zone may have to leave NATO's Nuclear Planning Group as well. Since they do not wish to be defended by nuclear weapons themselves, it might not be legitimate for them to participate in shaping the nuclear defences of other countries. On the other hand, in a nuclear war in Europe, the consequences would indeed be felt over the whole continent. Different countries would be differently affected, but there is no escape route for anyone. From that point of view, Nordic NATO members would still seem entitled to have a say in the formulation of nuclear strategies. The argument goes both ways.

Another implication of particular concern to NATO's nuclear weapon members is that a NWFZ could start a chain reaction that would shake the foundations of alliance nuclear strategy. Should Norway and Denmark drop out, the Netherlands may do the same, Belgium may follow suit, Greece may in any case drop out of the nuclear strategy, and so on. This is an important reason for US and British opposition to the zone proposals so far. It also explains much of the official West German reluctance, because it runs contrary to the German principle of non-singularity and the view that the nuclear burden should be shared among as many NATO members as possible. The more likely such a chain reaction is, the graver the Nordic reservations will seem, and the stronger the prospective sanctions against them, in terms of limitations on alliance participation and withdrawal of alliance support, will become. However, this might not be the case if the zone were to become part of a larger East–West rearrangement in Europe.

V. Geographical domain

In principle, the Nordic NWFZ proposals are *open-ended* in the sense that they allow for, invite or envisage more countries joining the zone as conditions become more propitious. They differ widely, however, concerning the initial domain of the zone.

As a first step in starting a process of denuclearization, it has been suggested that neutral countries such as Finland, Sweden, Switzerland and Yugoslavia could, unilaterally, reaffirm their nuclear weapon-free status and ask for affirmation of the non-use guarantees to which they are entitled [6]. More ambitiously, the starting-point could be Finland, Sweden, Norway and Denmark—including their territorial waters and airspace—although Danish security concerns are more tied to Central Europe than those of other Nordic countries, and therefore pose special problems (see section VII).[3] Iceland, another Nordic country, is in many ways less important for Western nuclear operations than Norway. But so far, the United States has not been willing to confirm that the Keflavik base is nuclear weapon-free, although the significance of the base for nuclear war-fighting purposes is commonly assumed to be on the decline. (Orion aircraft in the ASW role are becoming less dependent on nuclear depth charges.) The United States is also unwilling to confirm that nuclear weapons are not deployed at Thule and Söndre Strömfjord, Greenland; here, however, Denmark is in a rather good position to say whether they are or not, and except for the possibility of transit, these bases are virtually certain to be nuclear weapon-free.[4] For the rest of the Nordic area, the problem does not arise, essentially because the base policies of Denmark and Norway do not allow the stationing of foreign military personnel on their territories. For the zone to cover all Nordic territory—including Iceland, the Faeroe Islands and Greenland—a solution must therefore be found so that the Nordic countries can claim effective control over the entire area and reassure others that it is nuclear weapon-free. Of course, the nuclear weapon states must obligate themselves to respect the status of the zone, and thereby confirm that it is effectively free of nuclear weapons. The islands of Spitsbergen (belonging to Norway) and Aaland (belonging to Finland) have for several decades been demilitarized by treaty.

At sea, the territorial delimitation might follow the 12-mile rule. As for straits, the only strategically important strait in northern European waters is that leading into and out of the Baltic. Current Danish regulations demand that no more than three warships at a time can pass without special permission, and that submarines have to pass on the surface [8]. There is no special restriction on the passage of nuclear weapons. However, provisions for nuclear-armed warships may become desirable, depending on regulations to be agreed on nuclear weapon deployments in the Baltic Sea.

Territorial airspace is not clearly defined in international law. However, following the Warsaw Convention of 1929, it would reach as far up as modern planes can fly. Thus, intercontinental ballistic missiles are considered to travel in international outer space.

[3] Kekkonen's starting point was the 'continental areas' of Nordic countries, excluding Greenland and other islands as well as Iceland. A recent Soviet statement emphasized that Greenland ought to be part of the zone [7]. In a negotiation, the inclusion of Greenland might be traded against some similarly valuable Soviet concession.

[4] After the crash of a nuclear-armed B-52 aircraft near Thule in 1968, Denmark stressed that transit through the air territory of Greenland as well as storage of nuclear weapons on the island were prohibited.

VI. Transit provisions

The Treaty of Tlatelolco does not contain any provision regarding the transit of nuclear weapons. The Preparatory Commission for the Denuclearization of Latin America (COPREDAL) argued that it should be the prerogative of the territorial state, in the exercise of its sovereignty, to grant or deny permission for transit. In signing Additional Protocol II to the Treaty, the USA and France emphasized that each party to a nuclear weapon-free zone should retain exclusive legal competence to grant or deny transit. (This was motivated mainly by the use of the Panama Canal by the USA and other major powers.) In ratifying the same Protocol, the Soviet Union stated its objection to any such permission for transit.

For the Nordic countries, the transit of nuclear weapons mainly entails sea transit, except for Iceland (Keflavik). Even thus confined, it is a complex issue: it could involve a nuclear-armed ship showing the flag in a Danish harbour, ships participating in joint exercises, or an attack submarine calling at a Norwegian port for supplies or repair. Since large parts of the great power navies are equipped with nuclear weapons, it might be difficult for NATO members to prohibit all kinds of transit. An absolute prohibition could hamper joint military exercises to such an extent that allied support for Norway and Denmark would be seriously weakened. Such a prohibition would, moreover, be a rather one-sided concession on the part of Denmark, Norway and other Western powers.

The Soviet Baltic Fleet, and the significance of Soviet shipyards there, practically excludes prohibition of transit through the straits.

In the future, various kinds of air transit might also present great problems. Extended use by the United States of European airfields, including Danish and Norwegian, and extension of the Soviet air defence perimeter make the question more pertinent. In addition, there is the prospect of cruise missile transit, particularly the danger of cruise missile overflights of neutral air territory. Since Sweden and Finland can hardly defend themselves effectively against cruise missiles designed to penetrate Soviet air defences, political measures to counter this threat should be considered.

Under European 'buffer zone' conditions, the members of the zone cannot retain the competence to grant or deny transit as they please. The difference between a restrictive and a liberal practice would, under the circumstances, be too great to be left unregulated. Transits could, in theory, be so frequent that the basic non-stationing stand would be undermined. Transit provisions must therefore be negotiated.

Regardless of other zone provisions, overflights must be prohibited. In relation to cruise missiles, this could have some impact on the deployment of cruise missile carriers, making it less likely that the missiles would cross Nordic territory in time of war. SALT established functionally related observable differences (FRODs) for air-based systems, which made it possible to distinguish between B-52s with and without nuclear-loaded cruise missiles [9]. If such differences—related to cruise missile *carriers*—could also be established for sea-based systems, the obligations of nuclear weapon states could be made more precise, and the monitoring of adherence easier. However, this will be a very difficult task because some missiles will be submarine-based and others can be launched from standard tubes on a wide variety of surface vessels.

The missiles themselves have the same external physical characteristics whether they carry nuclear munitions or not, so it would seem that the overflight prohibition must apply to all cruise missiles, regardless of weapon load. However, since the deployment and movement of cruise missiles at sea are impossible to monitor with precision, and limitations extremely difficult to verify, a special treaty obligation to refrain from all plans and preparations that infringe on Nordic air territory might be the most that can be achieved. This obligation could be written into the Additional Protocol containing the guarantee for the zone, or into the provisions for deployment limitations in areas adjacent to the zone.

In general, the transit rules should be as strict as possible. However, it is even more difficult to prohibit in Europe activities which were not prohibited in the case of Latin America. The provisions regulating transit at sea have to be made both with regard to functional requirements for allied support to the NATO members in the area, and to the possibilities for verification. Formulation of the provisions would be complex but, provided that agreement is reached on certain political and military parameters, it might be relegated to a legal-technical operation of secondary importance.

VII. Deployment limitations in areas adjacent to the zone

In its foreign policy declaration of 18 March 1981, the Swedish government reiterated its long-standing view that a NWFZ agreement must include nuclear weapons "which are intended for targets within the zone, are stationed near the zone and have ranges of a scale which makes them best suited for targets within the Nordic area" [10]. Three months later, the Soviet Union stated its willingness to consider measures "applying to [Soviet] territory in the region adjoining the nuclear free zone in the North of Europe" [11]. Today, the viability of the zone proposals hinges very much on the prospects for deployment limitations in areas adjacent to the zone.

Figure 10.2. Target coverage of the Nordic area by NATO and Soviet missiles

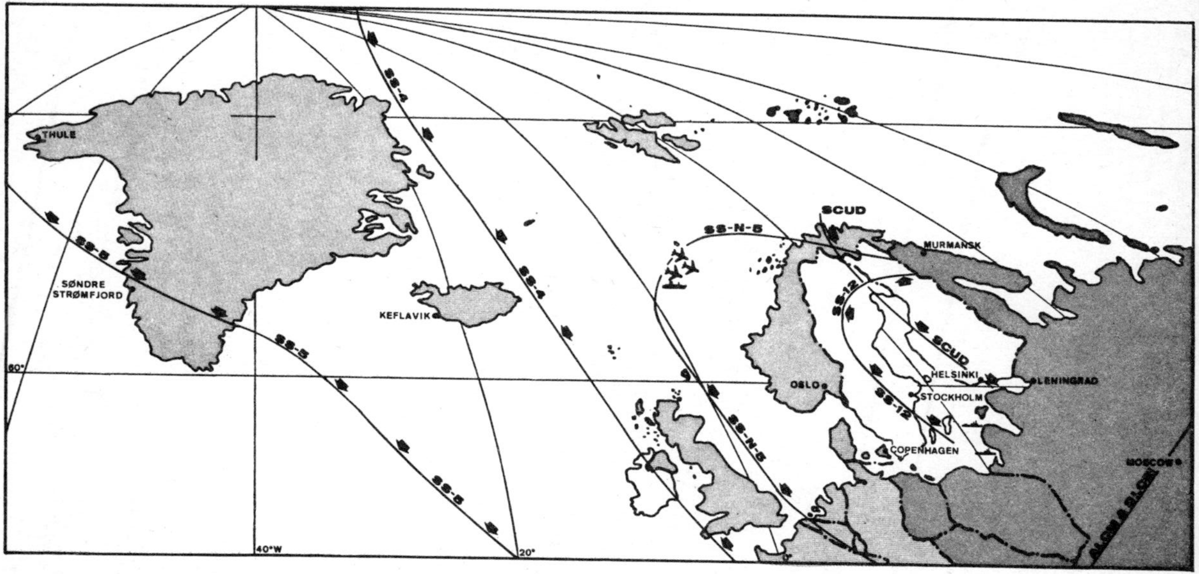

There are two main perspectives on the issue of deployment limitations. First, such limitations may be seen as a consequence of the guarantees for the zone. To the extent that nuclear weapons are unambiguously directed at targets within the zone—because of their geographical position, range or other indicators—they have to be removed; otherwise, they would constitute proof that the guarantees are fictitious. For example, the dozen or so 800-km range SS-12 Scaleboard missiles in the Leningrad military district are, in all likelihood, intended for interdiction strikes against Nordic targets, because they do not reach continental Western Europe.[5] The 350-km range Scud missiles in the same district are primarily intended for use against Nordic territory as well, in a tactical role. Scuds and Scaleboards belong to the standard Soviet weapon inventory at the Army and Front levels, respectively.

The elimination of weapons in this category is of special significance for the Nordic countries. The history of wars shows that belligerents usually do not surrender until all weapons have been used. Thus, weapons which can only be used against Nordic countries may, in an extreme situation, be used even against militarily insignificant targets on Nordic territory, as acts of terror. It is true that the elimination of these weapons would leave the nuclear powers with thousands of other weapons capable of striking targets in the Nordic area; but given that they can be used for a variety of important missions in other parts of the world, it is not certain that they would be used against a nuclear weapon-free zone. For the Nordic countries, the elimination of weapons without competing targets elsewhere is therefore more important than their relatively modest numbers would indicate.

However, few weapons can be used against Nordic countries only. Modern weapon systems are usable over varying distances and against different targets: they are becoming more mobile and more flexible, and can therefore meet a broad spectrum of military needs. It may therefore be more appropriate to seek deployment limitations as a matter of militarily significant confidence-building measures.

Regarding Soviet nuclear weapons this would, firstly, be a question of limitations in the Baltic Sea. Primary candidates for elimination are the six Soviet Golf II-class submarines, carrying altogether 18 SS-N-5 missiles with a range of 1 200 km. Other Soviet submarines in the Baltic are also likely to carry nuclear weapons—nuclear-tipped torpedoes as well as cruise missiles—essentially for use against sea targets, but to some extent suitable for land attack as well.[6] For the Soviet Union, the military

[5] The location of the SS-12 brigade is not known. However, even if deployed in the southern parts of the Leningrad military district, it would not reach FR Germany.

[6] The Soviet Whisky-class submarine which violated Swedish territorial waters and was stranded outside the town of Karlskrona in October 1981 seems to have had at least one nuclear-tipped torpedo on board.

usefulness of these weapons seems to be rather low, and they are increasingly obsolescent. The Western powers are not known to have any permanent deployment of nuclear-armed submarines in the Baltic. Therefore, it may not be unreasonable to ask for a total ban on submarine-based nuclear weapons in this area.

With a similar ban on surface-ship weapons, permanent deployment of nuclear weapons in the Baltic Sea would be prohibited altogether. Only transit to and from the bases and shipyards would be allowed. However, this is a tall order: actually, it is hard to imagine that total denuclearization of the Baltic Fleet could be achieved within the framework of a NWFZ arrangement. Alternatively, a partial prohibition of surface-ship weapons well suited for land attack might be considered. Or it could be left to the nuclear powers involved as a matter of unilateral restraint. After all, the guarantees for the zone raise expectations for the nuclear powers to show restraint in areas adjacent to it.

Secondly, some weapons deployed in the Leningrad military district may be withdrawn. Elimination of the 10 SS-5 launchers deployed on the Kola Peninsula would be a militarily significant confidence-building measure, as would the elimination of Scud and Scaleboard missiles in the same district. In the Northern Fleet, four Golf II-class submarines are candidates for removal. These diesel submarines are not SALT-accountable.

Toward the south, deployment limitations would apply first of all to the Schleswig-Holstein area, where nuclear weapons are known to be deployed in large numbers, but also to the southern shores of the Baltic Sea in general, affecting the German Democratic Republic and Poland as well. In relation to a zone confined to the Nordic area, this may raise great difficulties, because the predominant weapon carriers are multi-purpose aircraft,[7] and because they are organic parts of the Central European theatre. Deployment limitations in this area might therefore have to be discussed in terms of disengagement zones for Central Europe, geographically contiguous to a Nordic zone [6]. Limitations to the south seem, in other words, to depend on the establishment of militarily significant confidence-building measures in a wider European area.

This is of particular relevance to Denmark, which is responsible, together with FR Germany, for the defence of Denmark, Schleswig-Holstein (including Hamburg) and the Danish straits under a joint command established for this purpose (Commander Allied Forces Baltic Approaches,

[7] As in so many other arms control contexts, the aircraft sector poses very complex problems: deployment of aircraft is flexible; combat radii depend on many factors; many of the aircraft are dual-capable; and consequently, the variety of possible missions is large. To institute effective, unambiguous operational limitations on these forces therefore requires much ingenuity. Indeed, the complexity of the issue could make negotiations for a NWFZ long drawn out and, at worst, deadlocked.

COMBALTAP). Danish participation in a Nordic NWFZ could have a disruptive effect on this co-operation, as long as the West German forces operate on the basis of the NATO nuclear strategy. Therefore, measures which would increase the effectiveness and credibility of Western conventional defences, leading to a reduction of the role at present assigned to nuclear weapons, would facilitate Danish adherence to a NWFZ. And, even better, it would also be facilitated by mutual force reductions and the establishment of disengagement zones in Central Europe.

Preventing weapon *modernization*—that is, the substitution of new missiles and weapon carriers for old ones—is even more important than eliminating *existing* weapons of the types mentioned above. The new generations of weapons have improved war-fighting capabilities and appear as more threatening. Deployment limitations are therefore important, primarily for the options they block for the future and secondly for the weapons that would be removed.

VIII. Verification

In the Treaty of Tlatelolco, the provisions for verification are essentially geared to horizontal proliferation—the danger that states in the region might acquire nuclear weapons of their own. In the case of a Nordic NWFZ, the main verification requirements would relate to vertical proliferation, reassuring all states concerned that agreed restraints on the nuclear systems of established nuclear powers are observed.

For the members of the zone, IAEA safeguards and the treaty obligation to remain nuclear weapon-free should suffice. Guarantor states should not be given any special right to monitor or interfere with the activities of zone members. This would be politically unacceptable for the Nordic states and, for Sweden and Finland, incompatible with their policy of neutrality.

The main problem is to verify that the deployment limitations are observed. While this is difficult to discuss until the limitations are determined, verifiability is an important parameter of the elaboration of restrictions.

One thing seems obvious: since the Nordic countries themselves do not possess adequate technical means of verification, co-operation with the guarantor states is important. Being parties to the same arrangement prescribing limitations and restraints on both sides, the great powers must be presumed to watch each other with the means they have at hand. By establishing a joint commission where all states involved may raise matters for clarification or submit charges of violations, the members of the NWFZ would be in a position to draw upon the verification capabilities of the guarantor states. New issues could be referred to the same commission for clarification, that is, to a multilateral setting, thereby avoiding bilateral exchanges between one or more Nordic countries on the one hand, and a nuclear weapon state on the other.

However, this does not mean that all desirable limitations would be verifiable. Nor does it mean that a violation would necessarily be brought before the commission upon detection. A prohibition of submarine-based nuclear weapons in the Baltic may, for instance, be effectively verified in relation to ballistic missile-firing submarines, but probably not in relation to nuclear torpedoes, mines or cruise missiles that can be deployed on attack submarines. In the Norwegian Sea, airborne cruise missiles can be effectively monitored, but the movements of cruise missile-carrying submarines cannot. And even if violations are discovered by the great powers, they may not always find it in their interest to pass the information on to the members of the zone; the likelihood of bilateral horse-trading may not be high, but the possibility does exist.

Rigid demands for verification have often blocked the adoption of arms limitation measures for lack of trust, or have been used as a smokescreen for predominant interests in continued arms build-ups. In a period of high tension and low confidence, the great powers may once again rule out deployment limitations on the grounds of verification, contrary to the interests of the Nordic countries. For the latter, treaty obligations sustained by *some* possibilities of verification may be preferable to no limitation at all. However, since the limits are to be placed on the great powers, they cannot be implemented against their will.

IX. The European connection

A Nordic zone can be seen as a measure in its own right, although open-ended; as such, it may also be a first step towards a more comprehensive reduction of the numbers and roles of nuclear weapons in the European security system. Alternatively, it may be seen as an integral part of a broad European rearrangement, its fate being tied to developments on the larger European scene [12].

A number of European connections can also be envisaged following the first approach. One is obvious: the Geneva negotiations on theatre nuclear forces include such weapons as SS-5 missiles (on the Kola Peninsula) and Golf II missile-carrying submarines (in the Baltic and with the Northern Fleet), which might therefore be removed within that framework as well as in the zone context. Should the negotiations make progress and lead on to nuclear weapons with a shorter range than 1 000 km, limitations could be achieved on a broader range of weapons, including many of those deployed in areas adjacent to the zone. In the same manner, or by the adoption of militarily significant confidence-building measures as a follow-up to the Conference on Security and Co-operation in Europe (CSCE) in Madrid, the withdrawal of nuclear weapons from Central Europe could, furthermore, tie in with the Nordic NWFZ arrangement and provide a solution to the deployment limitation problem on the southern edge of the zone. In any case, a Nordic NWFZ initiative should be presented to all states participating in the CSCE, and their views and comments taken into account. This might facilitate the extension of the zone at a later stage, and encourage disengagement measures in other parts of Europe.

Alternatively, the Nordic countries might declare their willingness to establish a NWFZ in the Nordic area within the framework of a broader European arrangement, as an offer or contribution to arms reduction in a wider European domain. Following this approach, deployment limitations would not be sought as a consequence of the guarantees for the zone, or as a confidence-building measure attached to it; the road to containment and reduction of the threat of nuclear weapons to the Nordic area would go via nuclear disarmament in the wider European domain. Consequently, realization of a nuclear weapon-free zone in the North would depend on substantial progress in East–West disarmament talks. The matter would be left to the great powers, subject to their interests and priorities and, eventually, to their negligence.

X. Collateral measures

Other measures to strengthen the security of Nordic countries, and to stabilize relations between the great powers in northern Europe, can also be envisaged. They may be considered separately, or in conjunction with the zone idea, as collateral measures.

In relation to the zone proposal, ASW operations carried out or supported from Norwegian territory may merit particular attention. ASW systems would, no doubt, be important targets in a nuclear confrontation between the USA and the USSR, and might therefore draw Norwegian territory into the warfare. Today, the main ingredients of ASW activities from Norwegian territory are sonar arrays and Orion aircraft. Various types of Sound Surveillance Systems (SOSUS) are deployed in the area between Spitsbergen and Finnmark in northern Norway. They may have been deployed further east as well, together with other listening devices scattered throughout the Barents Sea. Norwegian Orion aircraft patrol as far east as 45°, that is, almost to Novaya Zemlya. The flights are co-ordinated with British Nimrod and US Orion aircraft from the Pitreavie Headquarters for the Northern Maritime Air Region in Scotland [13]. In recent years, US interest in improving its ASW capabilities in the Barents Sea seems to have grown. To some extent, this can be achieved by the introduction of new, self-contained technology that does not depend on local shore stations [14]. However, a much upgraded US capability in the Barents Sea is hardly going to leave Norwegian territory unaffected. Since Western ASW activities in the Barents Sea are aimed at the mainstay of Soviet retaliatory forces at sea (the West has no sea lanes to protect in the Barents Sea, so it can hardly be a question of tactical ASW), Soviet countermeasures must be expected. This would intensify the arms build-up in the area and may lead to a strengthened Soviet forward defence for the Kola base and its SSBN force—to the detriment of the security of Nordic states. Therefore, in the double interest of maintaining mutually assured destruction and enhancing the security of Norway and other Nordic countries, Norwegian-based ASW activity might, for instance, be limited to 24° East—following the self-imposed restriction not to allow allied air and naval units to cross that meridian over Norwegian territory. A restriction of this kind—which would not impede the protection of Atlantic sea lanes—would have the character of a collateral measure, and could be offered by Norway for consideration within the total context of the rights and obligations instituted by the zone arrangement.

Other confidence-building measures have been proposed and associated with the NWFZ idea as well, including a demilitarized area along the Norwegian–Soviet border, a somewhat broader area with agreed limits on military forces, and a political guarantee from Norway, Sweden and Finland that a conventional attack on Murmansk would not be allowed over their territories [15]. To the extent that the two major powers are still interested in maintaining mutual assured destruction, the vulnerability of the naval bases on the Kola Peninsula ought to be of concern for the United States as well. The more restraint the Western powers are willing to exercise and institute, the more far-reaching are the deployment limitations that can be asked of the Soviet Union and, consequently, the more substantial would be the restrictions on forces suited for attack on Nordic countries.

XI. Prospects and procedures

The Nordic NWFZ idea is of political interest because it has received remarkable public support in all the Nordic countries.

Norway and Denmark would, as a matter of course, have to consult with their allies on the drawing up of any zone arrangement. Equally obviously, the Nordic countries themselves must kick the ball off by taking a joint decision to initiate a process aiming at the establishment of a nuclear weapon-free zone in northern Europe. The decision might be taken at a meeting of Nordic foreign ministers. Iceland, a Nordic country and a regular participant in Nordic ministerial meetings, naturally ought to take part. Should Icelandic membership in the zone be considered premature, the meeting might underline the desirability of including Iceland at a later stage. Accordingly, it might also wish to emphasize that actions drawing Iceland deeper into Western nuclear strategy, as compensation for the denuclearization of Norway and Denmark, should be avoided.

Alternatively, the process could be initiated by co-ordinated declarations of all the countries to be included in the zone. One way or another, the constitution of the zone must be a Nordic initiative, even if it were to be presented as a Nordic offer to the great powers and other European states in the pursuit of arms reductions in the wider European domain. Otherwise, it would not carry much weight on the diplomatic scene.

Should the zone be seen as a measure in its own right, or as a first step towards a more comprehensive rearrangement in Europe, the Nordic NATO members would become involved in a sensitive balancing act between membership in the zone, on the one hand, and continued NATO membership on the other. On the one hand, they would have to meet the non-deployment demand and discontinue all preparations for transfer of nuclear weapons to their territories in time of crisis or war. On the other hand, the initial, rudimentary design must be of such a character, and have enough built-in flexibility, that the United States and NATO can accommodate the new conditions. If the United States declines to give guarantees for the zone, and if NATO balks at the alliance obligation to render support if limited to conventional means only, the zone is unlikely to be established.

It is hard to assess how difficult it would be to reconcile the two: no one knows precisely where the meeting points would be until negotiations have been held. There is no doubt that negotiations would raise great demands on the Nordic governments and foreign services in terms of both firmness and diplomatic flexibility.

XII. Concluding remarks

Deployment limitations in areas adjacent to the zone are crucial for the popular support and the ensuing vigour with which the Nordic countries will pursue the zone idea. With such limitations the significance of a NWFZ would be recognized even in peace-time.

In essence, the arrangement would be a militarily significant confidence-building measure, although more important politically than militarily. In time of crisis, it would function as an early-warning system—in the political rather than in the military sense. The procedural provisions regarding withdrawal or suspension of treaty obligations might play an important role here, and should be drafted so as to enhance the early-warning role. While the credibility of the guarantees in time of war would remain open to doubt, this question would become less important in the overall assessment of the merits of the arrangement because, to a large extent, the merits would be apparent in times of peace and crisis.

The composition of the limitation package is decisive for the consent and co-operation of the great powers. Above all, a balance must be struck which is compatible with the interests of the USA and the USSR. The elements to be balanced might be an unconditional Danish and Norwegian non-deployment stand, restraints on the movement of cruise missile carriers, and unilateral, collateral measures in return for Soviet arms reductions in the Baltic Sea and the Leningrad military district.

If there is no progress in the negotiations on long-range theatre nuclear forces at Geneva, it will be difficult to get any deployment limitations in the North: that would go against the general trend. If, on the other hand, the Geneva negotiations succeed, the need for separate limitations in northern Europe might gradually diminish, and the special restrictions to be attached to the zone made more manageable for the parties and the guarantor states to negotiate.

The case for disengagement in northern Europe is strong. The alternative to a zone arrangement—or to other arms limitation measures, for that matter—is not the *status quo* in northern Europe, but a big increase in military capabilities in the area. The latter would lead to increased tension and make the Nordic countries more vulnerable to great power confrontation elsewhere in the world, through the possible escalation and spread of armed conflict to the north of Europe.

The evolving growth and spread of more effective nuclear war-fighting weapons to the north of Europe underline the need for new measures to maintain a state of low tension in the area. At the same time it makes it more difficult to carry out such measures. It is not difficult to conceive of a NWFZ arrangement which would strengthen the security of the Nordic countries: the problem is to find a design which is acceptable to the major powers as well. This is but one example of a general, dialectic phenomenon in contemporary European affairs: while the arms race is more intense than ever before, at the same time public opinion against it is stronger than it has been for decades. It remains to be seen whether trends can be reversed; but the surging public interest in arms reduction, and in nuclear disarmament in particular, gives a glimpse of hope for the future.

References

1. President Kekkonen, Address at the Swedish Institute of International Affairs, Stockholm, 8 May 1978.
2. *Defense Week*, 8 June 1981.
3. George, J. L., 'US carriers—bold new strategy', *Navy International*, June 1981, pp. 330–35.
4. Prawitz, J., speech before the Swedish Kungliga Krigsvetenskapsakademien [Royal Military Science Academy], 13 March 1979.
5. Goldblat, J., *Agreements for Arms Control: A Critical Survey* (Taylor & Francis, London, 1982, Stockholm International Peace Research Institute).
6. Myrdal, A., in *Dynamics of European Nuclear Disarmament* (Spokesman, for European Nuclear Disarmament and the Bertrand Russell Peace Foundation, Nottingham, 1981).
7. *Information*, 7 July 1981.
8. Danish Royal Decree of 25 June 1951.
9. SIPRI, *World Armaments and Disarmament, SIPRI Yearbook 1980* (Taylor & Francis, London, 1980), chapter 7.
10. Protocol of the Parliamentary Foreign Affairs Committee, Stockholm, 18 March 1981.
11. *Suomen Sosialdemokraatti*, 26 June 1981.
12. Holst, J. J., 'The challenge from nuclear weapons and nuclear-weapon-free zones', *Bulletin of Peace Proposals*, No. 3, 1981.
13. Falchenberg, K., 'Ubåtsjakt i Norskehavet samordnas i Skottland', *Aftenposten* (Oslo), 5 February 1981.
14. SIPRI, *World Armaments and Disarmament, SIPRI Yearbook 1979* (Taylor & Francis, London, 1979), chapter 8.
15. Järvenpäa, P. and Ruhala, K., 'Arms Control in Northern Europe: Some Thoughts on a Nordic Nuclear-Free Zone', paper submitted to the Pugwash Symposium in Helsinki, April 1979 [unabridged Finnish version in *Ulkopolitiika*, No. 1, 1979].

11. The prohibition of inhumane weapons: new small arms ammunition

Square-bracketed numbers, thus [1], *refer to the list of references on page* 215.

I. Introduction

On 10 April 1981 the Convention on prohibitions or restrictions of use of certain conventional weapons was opened for signature at the United Nations in New York. The general form of the treaty is that of an 'umbrella' covering a number of protocols. The existing three protocols to the treaty are on landmines and booby-traps, incendiary weapons, and fragments not detectable by X-ray. (The texts of the Convention and the three protocols, and a discussion of them, are to be found in the *SIPRI Yearbook 1981*.)

The Convention includes no reference to the new small-calibre, high-velocity military rifle bullets which had been given high priority in the preparatory conferences, beginning with the conference of experts called by the International Committee of the Red Cross in 1973 [1]. Humanitarian concern about the effects of various bullet wounds goes back to the St Petersburg Declaration of 1868 (which outlawed small exploding and incendiary bullets) and the Hague Declaration of 1899 (which outlawed dumdum bullets).[1]

Between 1868 and 1899, most of the military powers introduced ammunition of about 7.62 mm calibre—a calibre which is still widely used today—with greatly increased velocity compared with the older ammunition. Scientific studies have shown this that ammunition could cause devastating injuries at ranges of up to several hundred metres, and fears were expressed about the casualties to be expected in future wars. However, these fears proved largely unfounded because the great range of the new ammunition (in excess of 1 000 m), together with the rapid rate-of-fire, changed the nature of warfare. The infantry and the cavalry could no longer cross the no-man's land between opposing forces, and warfare 'degenerated' from the heroic charges of a previous generation to the trench warfare of World War I.

In recent years there have been increasing demands to replace 7.62 mm full-power ammunition with shorter-range, lighter-weight ammunition and guns. Soviet-supplied forces have long relied primarily on a reduced-power ammunition with an effective range of about 300–400 m but with

[1] For a full discussion of this topic, see reference [2].

the same calibre. NATO countries were obliged by the United States to retain the full-power 7.62 mm ammunition and the heavier weapons it requires. Yet it was the USA itself which introduced lightweight ammunition and rifles in Viet Nam and subsequently throughout its forces. The US ammunition has a smaller calibre, 5.56 mm, and a higher velocity. There is considerable evidence that at ranges of up to several hundred metres (at which some 90 per cent of bullet wounds occur) this smaller calibre ammunition caused worse injuries than the larger calibre. The reason is that the bullet retarded more rapidly, giving up more energy in the wound; in addition it very frequently broke up.

It was these characteristics that led to the description of the new bullets (by the head of a Swedish government delegation, later foreign minister) as "the dumdum bullets of today". The Swedish government, with varying degrees of support from others, has since 1973 endeavoured to clarify the medical and technical issues involved by means of a series of international symposia on the wounding effects of modern assault rifle bullets. With the pressures of Viet Nam removed, the USA became more willing to co-operate in these efforts. However, when in 1979, at the first session of the UN Conference on these weapons, the Swedish government proposed that the UN itself should sponsor the fourth of the series of symposia, the proposal was strongly opposed by the USSR. A few months later the Soviet forces entering Afghanistan were seen to be equipped with new weapons, with 5.45 mm ammunition, which also appears to have worse wounding effects than the ammunition it replaces.

The Fourth International Symposium on Wound Ballistics was nevertheless held in Gothenburg, Sweden in September 1981.

The fact that the USA—which supplied many other countries with the new calibre ammunition—introduced weapons with worse wounding effects led to fears of a new and dangerous trend in the arms race: a trend which at first appeared to be confirmed by the new Soviet ammunition. These fears were reinforced by the opposition of the major powers to accepting restrictions on small arms ammunition within the framework of the UN Convention.

Further examination of the new ammunition now being introduced in NATO suggests that this unfortunate trend may have been stopped, perhaps as a result of the initiatives taken by the Swedish and other governments.

II. The M-16 and its ammunition

The US M-16 rifle is designed around a modified Remington .223-in high velocity hunting bullet: that is, in common terminology, it uses a .22 calibre rather than a .30 calibre round. Early versions had a rifling of 1 turn in 14 inches (1-in-350 mm). This resulted in a bullet, designated M-193, which was not very stable in flight, particularly in cold air. This meant that in Arctic conditions it proved almost impossible to hit the target; it also meant that the bullet tumbled very readily on impact, or even impacted sideways, causing severe injuries.

After a few years a rifling twist of 1-in-12 (1-in-305 mm) was introduced, resulting in a somewhat more stable bullet, and probably reducing the average severity of wounds.

These very technical considerations are important when comparing the M-16 with recent developments.

Clinical reports of injuries in Viet Nam indicated very severe injuries resulting from the M-16. Subsequent experimental studies, where bullets are fired into standard blocks of soap or gelatin, or into anaesthetized live animals, also gave evidence that the M-16 produced more severe effects than standard full-power (NATO) or reduced-power (Soviet) 7.62 mm ammunition. These experiments enabled another factor to be controlled, namely barrel wear. The high-powered ammunition of the M-16 resulted in a problem of barrel wear, which in turn also leads to unstable, and thereby inaccurate but potentially more dangerous, bullets.

The M-16 ammunition was criticized not only on humanitarian but also on military grounds. The lightweight bullet proved to have limited ability to penetrate foliage, steel plate on vehicles or helmets. The range was too limited for use in machine-guns. Thus, although many other armies have considered the need for smaller-calibre, lighter-weight weapons, there has been much dispute about the relative pros and cons of the various calibres. Thus, the USA went on to experiment with 6 mm ammunition for use in a light machine-gun, while the UK and the Federal Republic of Germany produced ammunition of less than 5 mm.

As a result of this confusion, NATO instituted extensive trials in order to arrive at a new common standard.

III. The new NATO 5.56 mm ammunition

On 28 October 1980, NATO approved a second standard ammunition. The Belgium SS-109 round selected has 5.56 mm calibre but it differs from the US M-193 5.56 mm round. It is designed to be fired from a gun with a rifling twist of 1-in-178 mm. The resulting rate of spin makes it stable in flight over a long distance, as well as being more stable on impact. It contains a hardened steel and lead core which penetrates steel plate more effectively than the larger-calibre NATO ammunition.

Figure 11.1. Cross-section of the cavity formed by standard military rifle bullets fired from 100 m into blocks of soap the approximate size of a human thigh

(A) Soviet 7.62 × 39 mm bullet; (B) NATO 7.62 × 51 mm bullet; (C) US 5.56 × 45 mm bullet.

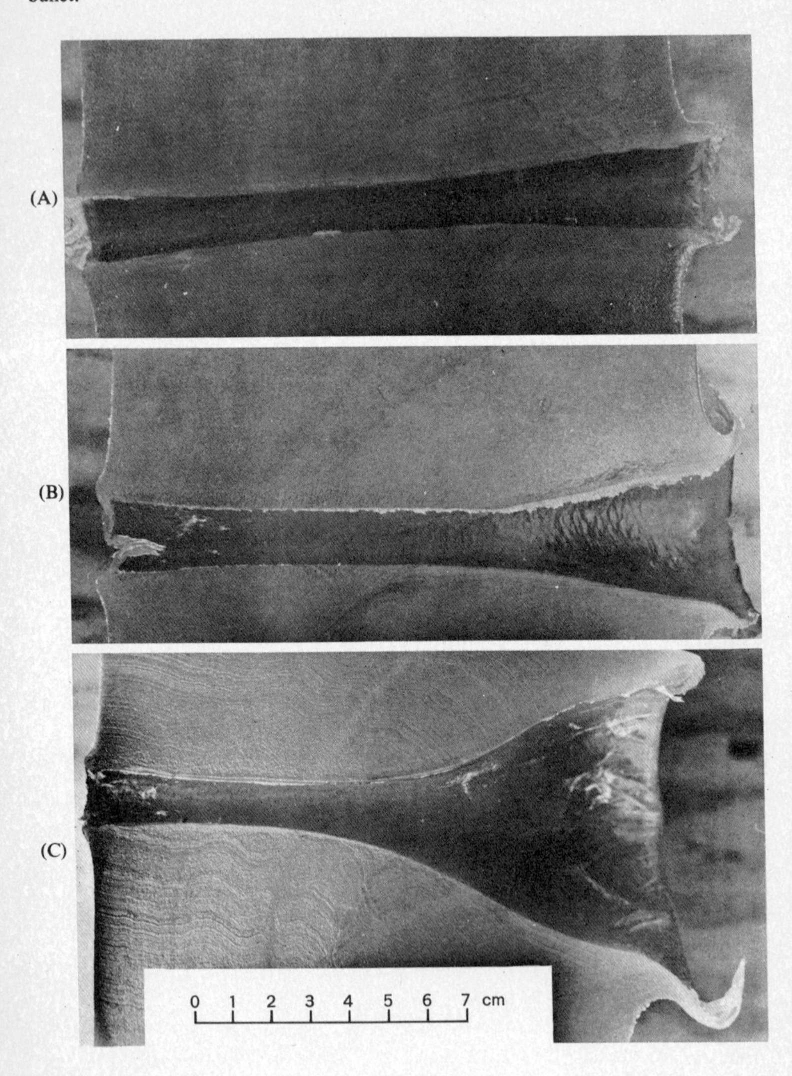

At the same time the designers appear to have taken account of the international concern about wounding effects and have produced a bullet which is said to be less severe than the M-193 round, having about the same effect as the 7.62 mm NATO ammunition (which is used in about 70 countries).

On the basis of the information published so far, it would seem that the Belgian designers have succeeded in producing a bullet which is militarily more effective (insofar as it can penetrate steel helmets at longer ranges) while at the same time being less objectionable from a humanitarian point of view.

A similar bullet, under the designation XM-855, is being manufactured by the United States. However, to be as effective, the barrels of the M-16 must be changed to ones with the 1-in-178 mm rifling. It seems likely that the USA will use up existing stocks of the M-193 ammunition and M-16 gun barrels and successively replace them with the new ammunition and new barrels or even new rifles, such as the M-16 PIP.

IV. The Soviet AK-74 and its ammunition

Not to be outdone, the Soviet Union has now introduced a smaller calibre assault rifle, the AK-74, and a light machine-gun, the RPKS-74. Both use 5.45 × 39 mm ammunition. The guns are very similar to those they replace, though somewhat improved.

Neither the Soviet ammunition nor the guns to fire it are substantially lighter than the ones they replace (they are already lighter than standard NATO 7.62 mm ammunition and rifles, and in fact slightly lighter than the Western 5.56 mm rounds and weapons). What, then, is the advantage of the new ammunition? The new *bullet* (not the *cartridge*) is about half the weight of the previous type. It is fired at higher velocity, giving a flatter trajectory and a greater accuracy at longer range. Further, the rifling of the gun (1-in-203 mm) makes the bullet very stable in flight. Even with the relatively high velocity (but which is less than that of the Western 5.56 mm bullets), the lightweight bullet has the lowest initial energy of any of the current military rifle bullets. This in turn raises questions about the lethality of the bullet.

Further examination of the bullet, however, shows that the Soviet designers have also considered this question. The bullet contains a mild-steel core surrounded by lead which forms a plug toward the tip of the bullet but fails to fill the tip. The centre of gravity is far to the rear ensuring that it will flip over when hitting the human body. It will thus very effectively deposit its energy in the body, causing an 'explosive type' wound, similar to that of the M-16. The steel jacket of the Soviet bullet makes it less liable to deformation or break-up than the US M-16 ammunition, while the mild-steel core gives it a greater ability to penetrate foliage or sheet metal.

There were at one time fears that the Soviet Union would adopt a 5.6 mm sporting bullet which causes more severe injuries than the M-16 ammunition. It now seems that they have not gone beyond the limits established by the USA.

V. The Fourth International Symposium on Wound Ballistics

The Fourth International Symposium on Wound Ballistics, held in Gothenburg, Sweden on 2–4 September 1981, covered a wide range of effects of wounding, predominantly wounding caused by small-calibre, high-velocity projectiles and mainly those from in-service rifles.[2] Work had previously been done on the more accurate detection of devitalized tissue and tissue which can be saved. Secondary effects of wounding had also been dealt with, for example in wounds caused by secondary fragments of bone which are driven many millimetres through the body after being struck by a bullet, thereby causing serious injury at some distance from the initial wound.

Eleven of the 34 papers presented at the 1981 Symposium dealt with the 'mechanical' effects of small-calibre, high-velocity bullets, in order to arrive at methods of measurement. Two approaches emerged from the Symposium experiments: many participants claimed that there were too many real-life variables to make it feasible to predict wound channels from experiments and that it was therefore impossible to create legislation; others argued that such evidence of the effects of missiles could establish some scale of the effects, with a view to restricting or prohibiting the use of those which caused the worst effects.

The Gothenburg Symposium was an effort on a technical level to reconcile the humanitarian and the military considerations for the design of small arms. What is needed now is the political will to prohibit by international law the ammunition which causes the most inhumane and indiscriminate wounds.

VI. Conclusions

The history of the M-16 and its successors enables a number of conclusions to be drawn. Once new standards are established by a great military

[2] The proceedings of the Third International Symposium on Wound Ballistics were published in reference [3]. It is understood that the proceedings of the Fourth Symposium will be published in the same journal.

power, they can spread rapidly throughout the world. The M-16 claimed victims not only in Indochina, but in Latin America, the Middle East and in Northern Ireland and Borneo.

Second, once again the Soviet Union demonstrated its determination to learn from US military experience, as well as its own.

Third, neither great power showed much respect for the rest of the international community or for efforts to restrain the development of new weapons by means of international humanitarian law.

Fourth, and more encouragingly, the new NATO standard does indicate that arms manufacturers and military authorities can be responsive not only to the 'technological imperative' but also to international public opinion.

It is to be hoped that the USA in its procurement policies, and the USSR in its development efforts, as well as other countries, will now pursue what is hopefully a tacit agreement to reverse the trend towards more inhumane small arms.

References

1. International Committee of the Red Cross, *Weapons that may Cause Unnecessary Suffering or have Indiscriminate Effects*, Report on the Work of Experts (ICRC, Geneva, 1973).
2. SIPRI, *Anti-Personnel Weapons* (Taylor & Francis, London, 1978).
3. *Acta Chirurgica Scandinavia*, Suppl. 489, 1979.

12. Disarmament at the 1981 UN General Assembly

Square-bracketed numbers, thus [1], *refer to the list of references on page* 219.

The 36th session of the UN General Assembly adopted over 50 resolutions on arms control issues. The most important resolutions are reviewed here.

I. Nuclear weapons

The Assembly reiterated its concern that nuclear weapon tests continued unabated against the wishes of the overwhelming majority of UN member states. It urged the UK, the USA and the USSR to resume their trilateral negotiations (interrupted in 1980) on a comprehensive test ban treaty (CTBT) and to bring them to an early conclusion [1]. It also requested the Committee on Disarmament (CD) to begin multilateral negotiations on a CTBT and, to this end, to establish a special working group, and called upon the UK, the USA and the USSR to halt, as a provisional measure, all their nuclear test explosions, either through a trilaterally agreed moratorium or through three unilateral moratoria [2].

The Assembly welcomed the commencement in November 1981 of US–Soviet negotiations on nuclear weapons, and urged that the strategic arms limitation process, begun by the conclusion of the SALT I agreements and the signature of the SALT II agreements, should continue and that the USA and the USSR, as signatory states, should refrain from any act which would defeat the object and the purpose of this process [3].

As in the previous years, appeals were made for nuclear weapon-free zones in different parts of the world. But the chances of having such zones, or 'zones of peace', actually set up, were not rated high: in Africa [4], in the Middle East [5, 6] in South Asia [7], or in the Indian Ocean [8], those being the regions specifically mentioned in the UN resolutions. The only positive development in this field that could be recorded by the General Assembly concerned Latin America: in November 1981, the United States deposited its instrument of ratification of Additional Protocol I of the Treaty of Tlatelolco, which prohibits nuclear weapons in Latin America [9]. Thereby the US Virgin Islands, the island of Puerto Rico and the US base of Guantanamo in Cuba have been submitted to the de-nuclearized regime.

As far as nuclear disarmament is concerned, most UN members consider the CD to be a suitable forum for the conduct of negotiations on the cessation of the production of nuclear weapons and on the reduction of their stockpiles, including their eventual elimination [10]. However, these negotiations are unlikely to be initiated in the foreseeable future in view of the strong objections on the part of the USA and its allies. On the other hand, the USSR and its allies are opposed to discussing a much less ambitious nuclear arms control measure—the cessation of the production of fissionable material for weapon purposes [11]—even though such a cut-off would certainly slow the manufacture of nuclear weapons and could perhaps even be a step toward ending such manufacture. The Soviet proposal for negotiating a convention to prohibit the production, stockpiling, deployment and use of neutron weapons was passed only by a narrow margin [12].

Reaffirming its call for effective international arrangements to assure non-nuclear weapon states against the use or threat of use of nuclear weapons [13], the Assembly suggested the conclusion of an international convention on the non-use of nuclear weapons in general, and declared that the use of such weapons would be a violation of the UN Charter and a crime against humanity [14]. This declaration was adopted by a considerable majority of states against the votes of the NATO countries, as was the declaration "on the prevention of nuclear catastrophe", which condemned the first use of nuclear weapons and the doctrines which envisage such use, although on this resolution a great number of states abstained [15]. The nuclear weapon states were requested to submit their views, proposals and practical suggestions for ensuring the prevention of nuclear war [16].

In a move related to nuclear arms control the Assembly referred to the 1981 Israeli attack against the Iraqi nuclear installations (already condemned by the IAEA General Conference) as an act directed against the IAEA and the nuclear safeguards regime [17]. It also expressed its "deep alarm" that Israel had the technical capability to manufacture nuclear weapons and possessed the means of delivery of such weapons (as stated in the report submitted by the Secretary-General) and requested the Security Council to prohibit all forms of nuclear co-operation with Israel [18].

The Assembly deplored the massive build-up of South Africa's military machine, including its "frenzied" acquisition of a nuclear weapon capability. The Security Council was asked to institute effective enforcement action against the South African regime so as to prevent it from endangering international peace and security through its acquisition of nuclear weapons [19].

II. Chemical weapons

While calling for the continuation of the negotiations on a convention prohibiting chemical weapons [20], the Assembly asked all states to refrain from any action which could impede such negotiations, and specifically to refrain from the production and deployment of binary and other "new" types of chemical weapons. It also requested that chemical weapons should not be stationed in those states where there are no such weapons at present [21].

Since the group of experts investigating the reports of the alleged use of chemical weapons (mainly US charges that the Soviet and Vietnamese forces had used toxic weapons in military operations in Afghanistan, Laos and Kampuchea) had not completed its work [22], the Assembly decided that this investigation should continue. Accordingly, the mandate of the group has been extended [23].

III. Other weapons

At the request of the Assembly a UN study will be carried out on all aspects of the conventional arms race and on disarmament relating to conventional weapons and armed forces. A group of qualified experts, appointed by the Secretary-General, is to be set up on a balanced geographical basis [24].

States have been asked to report annually (by 30 April) their military expenditures of the latest fiscal year for which data are available, making use of the standardized reporting instrument. The intention is to make these data an integral part of the regular UN statistical publications [25].

The Committee on Disarmament has been given the task of working out agreements to prevent the spread of the arms race to outer space, taking into account the Soviet draft treaty on the prohibition of the stationing of weapons of any kind in outer space [26]. Priority is to be given to negotiating an effective and verifiable ban on anti-satellite systems [27], a subject hitherto reserved for bilateral US–Soviet talks. Moreover, the Committee was requested to complete, during the first part of its session in 1982, the elaboration of a comprehensive programme of disarmament [28].

IV. Promotion of disarmament

The Assembly recommended that a world disarmament campaign should be launched, and that a pledging conference should be held at the United Nations to finance the campaign [29]. In addition, the Assembly expressed the view, with more than one-third of the membership abstaining, that a world-wide collection of signatures in support of measures to prevent nuclear war and to stop the arms race would be an important manifestation of the will of the world public and would contribute to the creation of a favourable climate for achieving progress in the field of disarmament. The Secretary-General was asked to work out the format and methods of conducting such an action under UN auspices [30].

V. Studies

The Assembly took note of several Secretary-General's reports on studies completed in 1981: on Israeli nuclear armament [31], on the relationship between disarmament and development [32], on the relationship between disarmament and international security [33], on confidence-building measures [34], on institutional arrangements relating to the process of disarmament [35], and on a world disarmament campaign [36].

(The study on the establishment of an international satellite monitoring agency (ISMA) [37] was submitted in August 1981 to the Preparatory Committee for the Second Special Session of the General Assembly devoted to disarmament.)

VI. Assessment

Although it adopted a record number of resolutions on arms control, the 1981 UN General Assembly did not break new ground in this field. The atmosphere of increased international tension was not conducive to progress. If anything, the sharp polemics between the USA and the USSR, characterized by mutual cold war accusations of aggressiveness and bad faith, have negatively affected the international arms control endeavours.

Most items, including the studies prepared by experts, have been referred for consideration to the Second Special Session of the General Assembly devoted to disarmament, which is due to take place from 7 June to 9 July 1982.

References

1. UN General Assembly resolution 36/85.
2. UN General Assembly resolution 36/84.
3. UN General Assembly resolution 36/97 I.
4. UN General Assembly resolution 36/86 B.
5. UN General Assembly resolution 36/87 A.
6. UN General Assembly resolution 36/87 B.
7. UN General Assembly resolution 38/88.
8. UN General Assembly resolution 36/90.
9. UN General Assembly resolution 36/83.
10. UN General Assembly resolution 36/92 E.

11. UN General Assembly resolution 36/97 G.
12. UN General Assembly resolution 36/92 K.
13. UN General Assembly resolution 36/95.
14. UN General Assembly resolution 36/92 I.
15. UN General Assembly resolution 36/100.
16. UN General Assembly resolution 36/81 B.
17. UN General Assembly resolution 36/27.
18. UN General Assembly resolution 36/98.
19. UN General Assembly resolution 36/86 A.
20. UN General Assembly resolution 36/96 A.
21. UN General Assembly resolution 36/96 B.
22. UN document A/36/613.
23. UN General Assembly resolution 36/96 C.
24. UN General Assembly resolution 36/97 A.
25. UN General Assembly resolution 36/82 B.
26. UN General Assembly resolution 36/99.
27. UN General Assembly resolution 36/97 C.
28. UN General Assembly resolution 36/92 F.
29. UN General Assembly resolution 36/92 C.
30. UN General Assembly resolution 36/92 J.
31. UN document A/36/431.
32. UN document A/36/356.
33. UN document A/36/597.
34. UN document A/36/474.
35. UN document A/36/392.
36. UN document A/36/458.
37. UN document A/AC.206/14.

13. Multilateral and bilateral arms control agreements[1]

The main undertakings which have been assumed by states in the arms control agreements concluded by 31 December 1981 include: (*a*) restrictions on nuclear weapon testing; (*b*) strategic arms limitations; (*c*) the non-proliferation of nuclear weapons; (*d*) the prohibition of non-nuclear weapons of mass destruction; (*e*) the demilitarization, denuclearization and other measures of restraint in certain environments or geographical areas; (*f*) the prevention of war; and (*g*) the humanitarian laws of war.

This chapter contains appropriately annotated summaries of these agreements.[2]

I. Restrictions on nuclear weapon testing

Treaty banning nuclear weapon tests in the atmosphere, in outer space and under water (Partial Test Ban Treaty—PTBT)

Signed at Moscow on 5 August 1963; entered into force on 10 October 1963

Prohibits the carrying out of any nuclear weapon test explosion or any other nuclear explosion: (*a*) in the atmosphere, beyond its limits, including outer space, or under water, including territorial waters or high seas; or (*b*) in any other environment if such explosion causes radioactive debris to be present outside the territorial limits of the state under whose jurisdiction or control the explosion is conducted.

Only three nuclear weapon powers—the UK, the USA and the USSR—are parties to the PTBT. China and France have refused to adhere to it, but France stopped atmospheric tests in 1975.

[1] The term 'arms control' is used here in a broad sense to denote measures intended to freeze, limit or abolish specific categories of weapons; to prevent certain military activities; to proscribe transfers of militarily important items; to reduce the risk of war; to constrain or prohibit the use of certain arms in war; or to build up confidence among states through greater openness in the military field. It thus includes measures of both arms limitation and disarmament.
[2] For the full texts of arms control agreements and the status of their implementation, see Goldblat, J., *Agreements for Arms Control: A Critical Survey* (Taylor & Francis, London, 1982, Stockholm International Peace Research Institute).

Treaty between the USA and the USSR on the limitation of underground nuclear weapon tests (Threshold Test Ban Treaty—TTBT)

Signed at Moscow on 3 July 1974; not in force by 31 December 1981

Prohibits from 31 March 1976 the carrying out of any underground nuclear weapon test having a yield exceeding 150 kt. Each party undertakes to limit the number of its underground nuclear weapon tests to a minimum. The provisions of the Treaty do not extend to underground nuclear explosions for peaceful purposes which are to be governed by a separate agreement. National technical means of verification are to be used to provide assurance of compliance, and a *protocol* to the Treaty specifies the data that have to be exchanged between the parties to ensure such verification.

Since the Treaty was not in force by 31 March 1976 (the agreed cut-off date for explosions above the established threshold) the parties stated that they would observe the limitation during the pre-ratification period.

Treaty between the USA and the USSR on underground nuclear explosions for peaceful purposes (Peaceful Nuclear Explosions Treaty—PNET)

Signed at Moscow and Washington on 28 May 1976; not in force by 31 December 1981

Prohibits the carrying out of any individual underground nuclear explosion for peaceful purposes, having a yield exceeding 150 kt, or any group explosion (consisting of two or more individual explosions) with an aggregate yield exceeding 1 500 kt. The Treaty governs all nuclear explosions carried out outside the weapon test sites after 31 March 1976. The question of carrying out individual explosions with a yield exceeding 150 kt will be considered at an appropriate time to be agreed. In addition to the use of national technical means of verification, the Treaty provides for access to sites of explosions in certain specified cases. A *protocol* to the Treaty sets forth operational arrangements for ensuring that no weapon-related benefits precluded by the TTBT are derived from peaceful nuclear explosions.

In 1977 the UK, the USA and the USSR started trilateral talks for the achievement of a comprehensive test ban treaty (CTBT). In 1980 these talks were adjourned *sine die*.

II. Strategic arms limitations

SALT I

Treaty between the USA and the USSR on the limitation of anti-ballistic missile systems (ABM Treaty)

Signed at Moscow on 26 May 1972; entered into force on 3 October 1972

Prohibits the deployment of ABM systems for the defence of the whole territory of the USA and the USSR or of an individual region, except as expressly permitted. Permitted

ABM deployments are limited to two areas in each country—one for the defence of the national capital, and the other for the defence of an intercontinental ballistic missile (ICBM) complex. No more than 100 ABM launchers and 100 ABM interceptor missiles may be deployed in each ABM deployment area. ABM radars should not exceed specified numbers and are subject to qualitative restrictions. National technical means of verification are to be used to provide assurance of compliance with the provisions of the Treaty.

The ABM Treaty is accompanied by *agreed interpretations* and *unilateral statements* made during the negotiations.

Protocol to the US–Soviet ABM Treaty

Signed at Moscow on 3 July 1974; entered into force on 25 May 1976

Provides that each party shall be limited to a single area for deployment of anti-ballistic missile systems or their components instead of two such areas as allowed by the ABM Treaty. Each party will have the right to dismantle or destroy its ABM system and the components thereof in the area where they were deployed at the time of the signing of the Protocol and to deploy an ABM system or its components in the alternative area permitted by the ABM Treaty, provided that, before starting construction, notification is given during the year beginning on 3 October 1977 and ending on 2 October 1978, or during any year which commences at five-year intervals thereafter, those being the years for periodic review of the ABM Treaty. This right may be exercised only once. The deployment of an ABM system within the area selected shall remain limited by the levels and other requirements established by the ABM Treaty.

Interim Agreement between the USA and the USSR on certain measures with respect to the limitation of strategic offensive arms

Signed at Moscow on 26 May 1972; entered into force on 3 October 1972

Provides for a freeze for a period of five years of the aggregate number of fixed land-based intercontinental ballistic missile (ICBM) launchers and ballistic missile launchers on modern submarines. The parties are free to choose the mix, except that conversion of land-based launchers for light ICBMs, or for ICBMs of older types, into land-based launchers for modern heavy ICBMs is prohibited. National technical means of verification are to be used to provide assurance of compliance with the provisions of the Agreement.

A *Protocol*, which is an integral part of the Interim Agreement, specifies that the USA may have not more than 710 ballistic missile launchers on submarines and 44 modern ballistic missile submarines, while the USSR may have not more than 950 ballistic missile launchers on submarines and 62 modern ballistic missile submarines. Up to those levels, additional ballistic missile launchers—in the USA over 656 launchers on nuclear-powered submarines and in the USSR over 740 launchers on nuclear-powered submarines, operational and under construction—may become operational as replacements for equal numbers of ballistic missile launchers of types deployed before 1964, or of ballistic missile launchers on older submarines.

The Interim Agreement is accompanied by agreed interpretations and unilateral statements made during the negotiations.

In September 1977 the USA and the USSR formally stated that, although the Interim Agreement was to expire on 3 October 1977, they intended to refrain from any actions incompatible with its provisions or with the goals of the ongoing talks on a new agreement.

Memorandum of Understanding between the USA and the USSR regarding the establishment of a Standing Consultative Commission on arms limitation

Signed at Geneva on 21 December 1972; entered into force on 21 December 1972

Establishes a Standing Consultative Commission (SCC) to promote the objectives and implementation of the provisions of the ABM Treaty and Interim Agreement of 26 May 1972, and of the Nuclear Accidents Agreement of 30 September 1971 (see below). Each government shall be represented by a commissioner and a deputy commissioner, assisted by such staff as it deems necessary. The Commission is to hold at least two sessions per year.

A *Protocol* establishing regulations governing the procedures and other relevant matters of the SCC was signed on 30 May 1973 and entered into force on the same day.

SALT II

Treaty between the USA and the USSR on the limitation of strategic offensive arms (SALT II Treaty)

Signed at Vienna on 18 June 1979; not in force by 31 December 1981

Sets, for both parties, an initial ceiling of 2 400 on intercontinental ballistic missile (ICBM) launchers, submarine-launched ballistic missile (SLBM) launchers, heavy bombers, and air-to-surface ballistic missiles (ASBMs) capable of a range in excess of 600 km. This ceiling will be lowered to 2 250 and the lowering must begin on 1 January 1981, while the dismantling or destruction of systems which exceed that number must be completed by 31 December 1981. A sublimit of 1 320 is imposed upon each party for the combined number of launchers of ICBMs and SLBMs equipped with multiple independently targetable re-entry vehicles (MIRVs), ASBMs equipped with MIRVs, and aeroplanes equipped for long-range (over 600 km) cruise missiles. Moreover, each party is limited to a total of 1 200 launchers of MIRVed ICBMs and SLBMs, and MIRVed ASBMs, and of this number no more than 820 may be launchers of MIRVed ICBMs. A freeze is introduced on the number of re-entry vehicles on current types of ICBMs, with a limit of 10 re-entry vehicles on the one new type of ICBM allowed each side, a limit of 14 re-entry vehicles on SLBMs and a limit of 10 re-entry vehicles on ASBMs. An average of 28 long-range air-launched cruise missiles (ALCMs) per heavy bomber is allowed, while current heavy bombers may carry no more than 20 ALCMs each. Ceilings are established on the throw-weight and launch-weight of light and heavy ICBMs. There are the following bans: on the testing and deployment of new types of ICBMs, with one exception for each side; on building additional fixed

ICBM launchers; on converting fixed light ICBM launchers into heavy ICBM launchers; on heavy mobile ICBMs, heavy SLBMs, and heavy ASBMs; on surface-ship ballistic missile launchers; on systems to launch missiles from the sea-bed or the beds of internal waters; as well as on systems for delivery of nuclear weapons from Earth orbit, including fractional orbital missiles. National technical means will be used to verify compliance. Any interference with such means of verification, or any deliberate concealment measures which impede verification, are prohibited. The Treaty is to remain in force until 31 December 1985.

Prior to the signing of the Treaty, on 16 June 1979, the USSR informed the USA that the Soviet 'Tu-22M' aircraft, called 'Backfire', is a medium-range bomber, and that the Soviet Union does not intend to give this bomber an intercontinental capability and will not increase its radius of action to enable it to strike targets on US territory. The USSR also pledged to limit the production of Backfire aircraft to the 1979 rate.

Protocol to the SALT II Treaty

Signed at Vienna on 18 June 1979; not in force by 31 December 1981

Bans until 31 December 1981: the deployment of mobile ICBM launchers or the flight-testing of ICBMs from such launchers; the deployment (but not the flight-testing) of long-range cruise missiles on sea-based or land-based launchers; the flight-testing of long-range cruise missiles with multiple warheads from sea-based or land-based launchers; and the flight-testing or deployment of ASBMs. The Protocol is an integral part of the Treaty.

The SALT II Treaty and the Protocol are accompanied by agreed statements and common understandings clarifying the obligations under particular articles.

In a **Memorandum of Understanding** the parties agreed on the numbers of strategic offensive arms in each of the 10 categories limited by the Treaty, as of 1 November 1978. In separate statements of data, each party declared that it possessed the stated number of strategic offensive arms subject to the Treaty limitations as of the date of signature of the Treaty.

Joint Statement by the USA and the USSR of principles and basic guidelines for subsequent negotiations on the limitation of strategic arms

Signed at Vienna on 18 June 1979

States that the parties will pursue the objectives of significant and substantial reductions in the numbers of strategic offensive arms, qualitative limitations on these arms, and resolution of the issues included in the Protocol to the SALT II Treaty. To supplement national technical means of verification, the parties may employ, as appropriate, co-operative measures.

As announced by the US Secretary of State, new strategic arms negotiations were to begin in the spring of 1982. In the meantime, on 30 November 1981, the United States and the Soviet Union started meeting in Geneva to conduct "intermediate nuclear force negotiations", as they were called by the USA, or "talks on the reduction of nuclear arms in Europe" as they were called by the USSR.

III. Non-proliferation of nuclear weapons

Treaty on the non-proliferation of nuclear weapons (NPT)

Signed at London, Moscow and Washington on 1 July 1968; entered into force on 5 March 1970

Prohibits the transfer by nuclear weapon states, to any recipient whatsoever, of nuclear weapons or other nuclear explosive devices or of control over them, as well as the assistance, encouragement or inducement of any non-nuclear weapon state to manufacture or otherwise acquire such weapons or devices. Prohibits the receipt by non-nuclear weapon states from any transferor whatsoever, as well as the manufacture or other acquisition by those states, of nuclear weapons or other nuclear explosive devices.

Non-nuclear weapon states undertake to conclude safeguards agreements with the International Atomic Energy Agency (IAEA) with a view to preventing diversion of nuclear energy from peaceful uses to nuclear weapons or other nuclear explosive devices.

The parties undertake to facilitate the exchange of equipment, materials and scientific and technological information for the peaceful uses of nuclear energy and to ensure that potential benefits from peaceful applications of nuclear explosions will be made available to non-nuclear weapon parties to the Treaty. They also undertake to pursue negotiations on effective measures relating to cessation of the nuclear arms race and to nuclear disarmament, and on a treaty on general and complete disarmament.

The structure and content of agreements between the IAEA and states required in connection with the NPT were agreed to in 1971. Pursuant to a safeguards agreement, the IAEA also concludes subsidiary arrangements which contain technical and operational details.

Of the five nuclear weapon powers, France and China have not adhered to the NPT. However, France stated that it would behave as a state adhering to the Treaty and that it would follow a policy of strengthening the safeguards relating to nuclear equipment, material and technology. Of the non-nuclear weapon states, India (not a signatory of the NPT) exploded in 1974 a nuclear device which it claimed to be for peaceful purposes.

In 1977 a group of major nuclear suppliers (the so-called London Club), comprising 15 countries, agreed on a set of guidelines for nuclear transfers.

Conferences of the parties to the NPT reviewing the implementation of the Treaty were held in 1975 and 1980.

Convention on the physical protection of nuclear material

Signed at Vienna and New York on 3 March 1980; not in force by 31 December 1981

Obliges the parties to ensure that, during international transport across their territory or on ships or planes under their jurisdiction, nuclear material for peaceful purposes as categorized in a special annex is protected at the agreed level. Storage of such material, incidental to international transport, must be within an area under constant surveillance. Robbery and embezzlement or extortion in relation to nuclear material, and acts without lawful authority involving nuclear material, are to be treated as punishable offences. "International nuclear transport" is defined as the carriage of a consignment of nuclear material by any means of transport intended to go beyond the territory of the state where the shipment originates.

UN Security Council Resolution on security assurances to non-nuclear weapon states

Adopted on 19 June 1968

Provides for immediate assistance by the UK, the USA and the USSR, in conformity with the UN Charter, to be given to any non-nuclear weapon state party to the NPT which is a victim of an act or an object of a threat of aggression in which nuclear weapons are used.

At the 1978 UN Special Session on Disarmament the USSR declared that it would never use nuclear weapons against those states which renounce the production and acquisition of such weapons and do not have them on their territories. The USA announced that it would not use nuclear weapons against any non-nuclear weapon state which is party to the NPT or any comparable internationally binding agreement not to acquire nuclear explosive devices, except in the case of an attack on the USA or its allies by a non-nuclear weapon state allied to or associated with a nuclear weapon state in carrying out or sustaining the attack. A similar statement was issued by the UK. Since then, the Committee on Disarmament has discussed ways of developing a uniform formula of security assurances to be incorporated in an international legal instrument.

IV. Prohibition of non-nuclear weapons of mass destruction

Convention on the prohibition of the development, production and stockpiling of bacteriological (biological) and toxin weapons and on their destruction (BW Convention)

Signed at London, Moscow and Washington on 10 April 1972; entered into force on 26 March 1975

Prohibits the development, production, stockpiling or acquisition by other means or retention of microbial or other biological agents, or toxins whatever their origin or method of production, of types and in quantities that have no justification for prophylactic, protective or other peaceful purposes, as well as weapons, equipment or means of delivery designed to use such agents or toxins for hostile purposes or in armed conflict. The destruction of the agents, toxins, weapons, equipment and means of delivery in the possession of the parties, or their diversion to peaceful purposes, should be effected not later than nine months after the entry into force of the Convention.

The 1980 Conference reviewing the operation of the BW Convention reaffirmed the comprehensive nature of the prohibitions under the BW Convention by stating that the language of the Convention fully covered all agents which could result from the application of such new techniques as the techniques for manipulation of molecules which form the genetic material of organisms.

The parties to the BW Convention recognized that the Convention was only a step towards an agreement effectively prohibiting also chemical weapons and providing for their destruction. Consequently, the prohibition of chemical means of warfare has been the subject of discussions in the Committee on Disarmament, as well as of bilateral talks between the USA and the USSR.

Convention on the prohibition of military or any other hostile use of environmental modification techniques (ENMOD Convention)

Signed at Geneva on 18 May 1977; entered into force on 5 October 1978

Prohibits military or any other hostile use of environmental modification techniques having widespread, long-lasting or severe effects as the means of destruction, damage or injury to states party to the Convention. The term "environmental modification techniques" refers to any technique for changing—through the deliberate manipulation of natural processes—the dynamics, composition or structure of the Earth, including its biota, lithosphere, hydrosphere and atmosphere, or of outer space.

The understandings reached during the negotiations, but not written into the Convention, define the terms "widespread", "long-lasting" and "severe".

Since 1979, the Committee on Disarmament has been discussing the prohibition of radiological weapons, defined as any device other than a nuclear explosive device, specifically designed to employ radioactive material by disseminating it to cause destruction, damage or injury by means of the radiation produced by the decay of such material, as well as any radioactive material, other than that produced by a nuclear explosive device, specifically designed for such use.

V. Demilitarization, denuclearization and other measures of restraint in certain environments or geographical areas

Antarctic Treaty

Signed at Washington on 1 December 1959; entered into force on 23 June 1961

Declares the Antarctic an area to be used exclusively for peaceful purposes. Prohibits any measure of a military nature in the Antarctic, such as the establishment of military bases and fortifications, and the carrying out of military manoeuvres or the testing of any type of weapon. Bans any nuclear explosion as well as the disposal of radioactive waste material in Antarctica, subject to possible future international agreements on these subjects.

Representatives of the contracting parties meet at regular intervals to exchange information and consult each other on matters of common interest pertaining to Antarctica, as well as to recommend to their governments measures in furtherance of the principles and objectives of the Treaty.

Treaty on principles governing the activities of states in the exploration and use of outer space, including the Moon and other celestial bodies (Outer Space Treaty)

Signed at London, Moscow and Washington on 27 January 1967; entered into force on 10 October 1967

Prohibits the placing in orbit around the Earth of any objects carrying nuclear weapons or any other kinds of weapons of mass destruction, the installation of such weapons on celestial bodies, or the stationing of them in outer space in any other manner. The establishment of military bases, installations and fortifications, the testing of any type of weapons and the conduct of military manoeuvres on celestial bodies are also forbidden.

A separate **Agreement governing the activities of states on the Moon and other celestial bodies** was opened for signature on 18 December 1979. By 31 December 1981 it was not yet in force.

In 1981 the Soviet Union proposed a treaty which would prohibit the stationing of weapons of any kind in outer space, including stationing on reusable manned space vehicles.

Treaty for the prohibition of nuclear weapons in Latin America (Treaty of Tlatelolco)

Signed at Mexico City on 14 February 1967; entered into force on 22 April 1968

Prohibits the testing, use, manufacture, production or acquisition by any means, as well as the receipt, storage, installation, deployment and any form of possession of any nuclear weapons by Latin American countries.

The parties should conclude agreements with the IAEA for the application of safeguards to their nuclear activities.

Under *Additional Protocol I*, annexed to the Treaty, the extra-continental or continental states which, *de jure* or *de facto*, are internationally responsible for territories lying within the limits of the geographical zone established by the Treaty (France, the Netherlands, the UK and the USA), undertake to apply the statute of military denuclearization, as defined in the Treaty, to such territories.

Under *Additional Protocol II*, annexed to the Treaty, the nuclear weapon states undertake to respect the statute of military denuclearization of Latin America, as defined in the Treaty, and not to contribute to acts involving a violation of the Treaty, nor to use or threaten to use nuclear weapons against the parties to the Treaty.

Treaty on the prohibition of the emplacement of nuclear weapons and other weapons of mass destruction on the sea-bed and the ocean floor and in the subsoil thereof (Sea-Bed Treaty)

Signed at London, Moscow and Washington on 11 February 1971; entered into force on 18 May 1972

Prohibits emplanting or emplacing on the sea-bed and the ocean floor and in the subsoil thereof beyond the outer limit of a sea-bed zone (coterminous with the 12-mile outer limit of the zone referred to in the 1958 Geneva Convention on the Territorial Sea and the Contiguous Zone) any nuclear weapons or any other types of weapons of mass destruction as well as structures, launching installations or any other facilities specifically designed for storing, testing or using such weapons.

The 1979 SALT II Treaty extended, for the USA and the USSR, the ban on military activities in the sea-bed environment. It prohibits the development, testing or deployment of fixed ballistic or cruise missile launchers for emplacement on the ocean floor, on the sea-bed, or on the beds of internal waters and inland waters, or in the subsoil thereof, or mobile launchers of such missiles, which move only in contact with the ocean floor, the sea-bed, or the beds of internal waters and inland waters, or missiles for such launchers.

Document on confidence-building measures and certain aspects of security and disarmament, included in the Final Act of the Conference on Security and Co-operation in Europe (CSCE)

Signed at Helsinki on 1 August 1975

Provides for notification of major military manoeuvres in Europe to be given at least 21 days in advance or, in the case of a manoeuvre arranged at shorter notice, at the earliest possible opportunity prior to its starting date. The term "major" means that at least 25 000 troops are involved. States may invite observers to attend the manoeuvres.

At the follow-up meeting of the CSCE in 1980–81, proposals were made for mandatory notification of military manoeuvres and movements with fewer than 25 000 men, for setting an earlier date for notification, and for providing observers with substantive information.

Since 1973, talks on the reduction of forces and armaments in Central Europe have been held in Vienna.

VI. Prevention of war

Memorandum of Understanding between the USA and the USSR regarding the establishment of a direct communications link ('Hot Line' Agreement)

Signed at Geneva on 20 June 1963; entered into force on 20 June 1963

Establishes a direct communications link between the governments of the USA and the USSR for use in time of emergency. An annex attached to the Memorandum provides for two circuits, a duplex wire telegraph circuit and a duplex radio telegraph circuit, as well as two terminal points with telegraph–teleprinter equipment between which communications are to be exchanged.

An agreement signed on 30 September 1971 improved the reliability of the US–Soviet Hot Line by providing for the establishment of two satellite communications circuits between the USA and the USSR, with a system of multiple terminals in each country.

Direct communications links have also been established between France and the USSR, as well as between the UK and the USSR, following the agreements concluded in 1966 and 1967, respectively.

Agreement between the USA and the USSR on measures to reduce the risk of outbreak of nuclear war ('Nuclear Accidents' Agreement)

Signed at Washington on 30 September 1971; entered into force on 30 September 1971

Provides for immediate notification in the event of an accidental, unauthorized incident involving a possible detonation of a nuclear weapon (the party whose nuclear weapon is involved should take necessary measures to render harmless or destroy such weapon); immediate notification in the event of detection by missile warning systems of unidentified objects, or in the event of signs of interference with these systems or with related communications facilities; and advance notification of planned missile launches extending beyond the national territory in the direction of the other party.

The 1979 SALT II Treaty extended the obligations of the parties with regard to advance notification of missile launches. All planned multiple launches (that is, those which would result in two or more ICBMs being in flight at the same time), even if the planned trajectories were to be entirely within a party's national territory, would have to be notified.

The French–Soviet and British–Soviet Nuclear accidents Agreements, concluded in 1976 and 1977, respectively, are patterned after the US–Soviet Agreement.

Agreement between the USA and the USSR on the prevention of incidents on and over the high seas

Signed at Moscow on 25 May 1972; entered into force on 25 May 1972

Provides for measures to assure the safety of navigation of the ships of the armed forces of the USA and the USSR on the high seas and flight of their military aircraft over the high seas, including rules of conduct for ships engaged in surveillance of other ships as well as ships engaged in launching or landing aircraft. The parties also undertake to give notification of actions on the high seas which represent a danger to navigation or to aircraft in flight, and to exchange information concerning instances of collisions, instances which result in damage, or other incidents at sea between their ships and aircraft.

In a *Protocol* signed in 1973, the parties undertook that their ships and aircraft should not make simulated attacks by aiming guns, missile launchers, torpedo tubes and other weapons at non-military ships of the other party, nor launch nor drop any objects near non-military ships of the other party in such a manner as to be hazardous to these ships or to constitute a hazard to navigation.

Agreement between the USA and the USSR on the prevention of nuclear war

Signed at Washington on 22 June 1973; entered into force on 22 June 1973

Provides that the parties will act in such a manner as to exclude the outbreak of nuclear war between them and between either of the parties and other countries. Each party will refrain from the threat or use of force against the other party, against the allies of the other party and against other countries in circumstances which may endanger international peace and security. If at any time relations between the parties or between either party and other countries appear to involve the risk of a nuclear conflict, or if relations between countries not parties to this Agreement appear to involve the risk of nuclear war between the USSR and the USA or between either party and other countries, the Soviet Union and the United States, acting in accordance with the provisions of this Agreement, shall immediately enter into urgent consultations with each other and make every effort to avert this risk.

VII. The humanitarian laws of war

Protocol for the prohibition of the use in war of asphyxiating, poisonous or other gases, and of bacteriological methods of warfare (Geneva Protocol)

Signed at Geneva on 17 June 1925; entered into force on 8 February 1928

Declares that the parties agree to be bound as between themselves by the above prohibition, which should be universally accepted as part of international law, binding alike the conscience and the practice of nations.

Convention on the prevention and punishment of the crime of genocide (Genocide Convention)

Adopted at Paris by the UN General Assembly on 9 December 1948; entered into force on 12 January 1951

Declares genocide, defined as the commission of acts intended to destroy, in whole or in part, a national, ethnic, racial or religious group, as such, to be a punishable crime.

Conventions for the protection of war victims (Geneva Conventions)

Signed at Geneva on 12 August 1949; entered into force on 21 October 1950

Convention I provides for the amelioration of the condition of the wounded and sick in armed forces in the field.
Convention II provides for the amelioration of the condition of the wounded, sick and shipwrecked members of armed forces at sea.
Convention III relates to the treatment of prisoners of war.
Convention IV relates to the protection of civilian persons in time of war.

Protocol (I) Additional to the 1949 Geneva Conventions

Signed at Bern on 12 December 1977; entered into force on 7 December 1978

Relates to the protection of victims of international armed conflicts.

Reiterates the rule of international law that the right of the parties to an armed conflict to choose methods or means of warfare is not unlimited, and that it is prohibited to use weapons and methods of war that cause superfluous injury or unnecessary suffering. Expands the existing prohibition against indiscriminate attacks to cover attacks by bombardment of cities or other areas containing a similar concentration of civilians or civilian objects. Dams, dykes and nuclear electric power generating stations are placed under special protection. There is also a prohibition to attack, by any means, localities declared as non-defended, or to extend military operations to zones on which the parties conferred by agreement the status of demilitarized zone. Reprisals against the civilian population are forbidden. Guerrilla fighters are accorded the right to prisoner-of-war status if they belong to organized units subject to an internal disciplinary system and under a command responsible to the party concerned.

Protocol (II) Additional to the 1949 Geneva Conventions

Signed at Bern on 12 December 1977; entered into force on 7 December 1978

Relates to the protection of victims of non-international conflicts.

Prescribes humane treatment of all the persons involved in such conflicts, care for the wounded, sick and shipwrecked, as well as protection of civilians against the dangers arising from military operations.

Convention on the prohibitions or restrictions on the use of certain conventional weapons which may be deemed to be excessively injurious or to have indiscriminate effects

Signed at New York on 10 April 1981; not in force by 31 December 1981

The Convention is an 'umbrella treaty', under which specific agreements can be concluded in the form of protocols.

Protocol I prohibits the use of weapons intended to injure by fragments which are not detectable in the human body by X-rays.

Protocol II prohibits or restricts the use of mines, booby-traps and similar devices.

Protocol III prohibits or restricts the use of incendiary weapons.

INDEX

Page numbers followed by *n* refer to footnote references.

World Armaments and Disarmament, SIPRI Yearbook 1982: Contents List

Those chapters and appendices marked with an asterisk are included in this paperback.

Recent and Forthcoming SIPRI Books